RAISING POULTRY THE MODERN WAY

by

Leonard S. Mercia

GARDEN WAY PUBLISHING

Charlotte, Vermont 05445

RAISING POULTRY THE MODERN WAY

Published in the United States by Garden Way Publishing, Charlotte, Vermont 05445.

Library of Congress Catalog Card Number: 74-75463
International Standard Book Number: 0-88266-058-6

CONTENTS

RAISING POULTRY THE MODERN WAY

INTRODUCTION

The incentive to write this book resulted from the many recent requests for information about the small poultry flock. During the nearly 20 years the author has been associated with poultry extension education never has there been the interest in the small family flock that currently exists.

For years we have witnessed the development of a large, highly automated, and specialized poultry industry. This revolution has taken place at the expense of the small poultry flock because of diminishing profits and problems of marketing for small flock owners. Now, very suddenly, there has developed a renewed interest in raising a few chickens, turkeys or waterfowl. Why this sudden turnabout?

The desire to produce fresh eggs and poultry for the family probably ranks high among the reasons for starting a small flock. However, a few birds may also provide family members with an excellent chore responsibility, a hobby, or even a source of limited income. Whatever the reason, success in such a venture requires a knowledge of poultry husbandry and many other related subjects.

The need exists for a handy reference manual which contains complete and practical information for the small poultryman or prospective poultryman. This book is written with that need in mind. It treats practically all phases of poultry production and processing and includes sections on chickens, turkeys and waterfowl.

One chapter attempts to help answer the question, "Should I Raise A Poultry Flock?" and includes a discussion of the possible types of poultry projects and some of the advantages and disadvantages of each. The section which will probably interest the larger number of readers is the one on the laying flock, since this seems to be the project that is of interest to most people. Much of the chapter centers around questions the author is frequently asked. It therefore covers topics such as types and sources of stock, management and feeding practices, housing and equipment needs. It contains plans for small poultry houses, poultry furniture, a small incubator and others. One section is devoted to meat bird production. Other sections treat turkey and waterfowl production. Information is presented on such skills as culling, debeaking, caponizing, egg care, egg grading and poultry processing. To enhance the value of the book as a handy reference, it includes lists of additional reading materials, sources of supplies and equipment, as well as the addresses of poultry specialists at the State Universities and the State Poultry Diagnostic Laboratories.

Since the book does include information on more than one species, and many of the subjects discussed required references to other subjects, some repetition was unadvoidable. Then too, it was felt that repetition might be helpful to those readers who have an interest in only one section of the book or who have time to read only one section at a time. However, cross references are used to prevent as much duplication as possible.

Although intended primarily as a guide and reference for prospective poultrymen and small flock owners, *Raising Poultry the Modern Way* does also have value as a text or supplement for practical poultry production courses or science projects in the schools.

Leonard S. Mercia

1

SHOULD I RAISE POULTRY?

Today, the care of a small poultry flock can be a rewarding experience as a hobby, to help fill the family refrigerator or freezer, or to sell. Modern poultry keeping can also be an enriching experience. With good husbandry and flock management the poultry project can be a source of efficiently produced food. Commercially produced poultry feeds can be supplemented to some extent with table scraps, green garden crops, and lawn clippings — up to one third of the nutritional requirements of the flock. The selection of the right stock for your particular type of project will help provide optimum results, whether it involves laying birds, broilers, roasters, capons, turkeys, waterfowl, or others. Moreover, surplus eggs and poultry may be sold at a healthy profit.

We are currently experiencing a resurgence of interest in the production of food at home. Often the responsibility for the care of the poultry project is given to the younger members of the family. This can be a well-rounding experience for young people, fostering a basic knowledge of such things as egg care, incubation, feeding and management of growing birds and, if properly undertaken, the development of such important skills as record-keeping, and money management.

A valuable by-product of the poultry project which is frequently overlooked is the manure. Poultry manure is high in fertilizer value and is an excellent fertilizer for the garden or a valuable addition to the compost pile.

The modern commercial poultry industry is perhaps best characterized as being a very highly specialized, highly commercialized, agricultural industry. It is no longer a small business on many farms, but a large business on fewer and fewer farms. The keys to success in the commercial poultry business are frequently bigness and efficiency. The necessary capital investment is high; there are some pitfalls in the business, and a profitable market for large volume production is nearly essential if the enterprise is to be successful.

Small poultry enterprises with a few hundred or a few thousand birds can be profitable if the products can be marketed at a substantial margin direct to stores, restaurants, hospitals, institutional users, or direct to consumers. Even the small family flock can return a profit when products are marketed direct to the housewife at a substantial markup. These outlets, too, are not always free of problems. Competition for markets is keen, and profit margins sometimes erode. Moreover, there is the problem of providing a volume of sizes, quality and service that is often difficult for the small producer who sells to the wholesale outlets.

If you plan to sell your products and make a profit, you need to study the market potential in your area carefully. Without profitable markets any poultry enterprise is almost certainly doomed to failure.

Before starting a poultry flock, check the local zoning ordinances. Animal projects are not permitted in certain residential areas. A possible problem might involve your neighbors. Poultry flocks can bring about such problems as odors, flies, rats or mice. There is the possibility of complaints about noise, especially if there are adult males in the chicken flock. Nothing manages to anger neighbors more than the early morning crowing of a rooster. Good management will prevent many of these nuisances, but it is well to research thoroughly all possible problems before embarking upon your own poultry project.

Should you decide to raise poultry, it is, in most instances, a seven-day-a-week job. There are the daily management chores to be done — feeding, watering, collecting eggs and checking the birds.

Types of Poultry Projects

the laying flock

The Laying Flock is the most popular type of poultry project. Good stock properly managed will lay eggs throughout the year. Laying birds should start laying at 22-24 weeks of age. The length of their laying cycle is normally 12 to 14 months. Any stock that is bred for high egg production is suitable for the laying flock. A good laying bird, under good management conditions, should lay 19-20 dozen eggs during a laying cycle. Then too, there is the possibility of an occasional meal of fowl when poor producers are culled from the flock.

meat chickens

Specialty meat-type birds such as capons and roasters are particularly suited to small flock enterprises. Broilers, as a general rule, cannot be grown on the small farm as economically as they can be grown commercially. Broilers are young chickens used for frying, broiling or roasting and are usually about eight weeks of age when processed. Commercial broilers are scientifically bred to produce meat efficiently. Very few broilers are now produced as a by-product of egg production.

Roasters are 3-5 month old chickens that weigh 5 pounds or more. Capons are castrated males grown for 6-7 months to weights of 7 pounds or more. The costs of raising these chickens to heavier weights are considerably higher than the costs of raising broilers. These tender-meated, well-finished birds are, however, sought after and demand a much higher price-per-pound in the market.

turkeys

Turkey production makes an excellent project for fun or profit. Although it was once thought to be hazardous and difficult to raise turkeys, this is no longer true. Thanks to the available modern-day drugs and management know-how, the turkey enterprise is not now a particularly hazardous one. Still, there are certain precautions which have to be observed. Blackhead disease is a possible problem, and turkeys should be reared separately from chickens to avoid problems with this disease. It is possible to raise turkeys either in confinement or on range.

A turkey project can be an excellent one for the young or old. It can provide an excellent responsibility and learning experience for young people, and with good stock, good management and marketing, a profit can be realized. Turkeys can be raised in virtually any type of climate successfully if they receive the proper management and nutrition and are protected against diseases, predatory animals and exposure to extreme weather conditions.

Turkey production has the added advantage of being a relatively short-term project since it requires only about 5 months to prepare for the poults, grow them, process and market them and clean the facilities in preparation for the next flock.

Fresh native turkey, properly dressed, is hard to beat for juiciness and flavor, so it is a natural for the production of food for home consumption.

waterfowl

Like other phases of the poultry industry, there are farms that have specialized in commercial production of ducks and geese. There are also many small flocks raised to occupy the farm pond or as a hobby or sideline to other farm enterprises. Ducks and geese make an excellent hobby, and many are raised for exhibition purposes or multiplied for sale as ducklings or goslings.

Ducks and geese, which are well-finished, are a nutritious and tasty food. Some of the breeds of ducks, surprisingly enough, surpass some chickens in egg-laying ability.

Rearing waterfowl is relatively easy. The young are quite hardy, and by following reasonable management and feeding programs, such a project can be a successful and profitable one. Ducks and especially geese are great foragers and can pick up a great deal of their nutritive requirements during the warm weather months in the form of green feed, insects and worms. In confinement they can cause damp litter conditions, unless special waterers and management procedures are observed.

2

POULTRY HOUSING
AND EQUIPMENT

The type of poultry house will vary with the type of project and how you plan to get started. For example, baby chicks and turkey poults require tighter, more comfortable housing than do older birds. If you plan to begin the project with started birds, then the house may not have to be as tight or as well-insulated. In warm climates the house may have open sides enclosed with poultry netting. It may be equipped with adjustable awnings for use during inclement weather. Waterfowl require very little in the way of housing as adults. However, young ducklings and goslings, though relatively hardy, will need good housing during the early brooding period.

The type of house will also depend, to a certain extent, upon the management system you use. Some differences in design and size might be needed to accommodate different types of equipment and furniture. Cages may be utilized for laying flocks, and thus the building construction and design must be considered before building to insure that the equipment fits into the building. Material-handling chores such as feeding, cleaning, watering and gathering eggs should also be considered when planning the poultry house.

The poultry house need not be elaborate or expensive. An old building can frequently be remodelled to accommodate a small flock of poultry. Sometimes a pen can be built within a large existing structure that will function quite well for small flocks. Remodelling of old buildings for large production units is seldom advisable. Though some old structures are suitable for the production of broilers, pullets, or turkeys, most are not readily adapted to the efficient handling of birds. While material-handling chores for young growing birds are not so critical, it is an important consideration for large laying projects.

There are basically two types of poultry housing. One type is designed for floor management systems, the other for cage systems.

Floor housing, sometimes called litter or loose housing, allows the birds free access to the poultry house or pens within the house. The advantages of the floor system are that it permits flexibility in the use of the house. It can be adapted for brooding and growing or managing most any type of poultry flock including broilers, roasters, capons, layers, turkeys or waterflowl. This system is well suited for the small farm or family flock, and most of our consideration will be given to this type of housing.

A word about cages. This is the system most frequently used by commercial egg producers today. There are several reasons for its popularity. Floor space requirement per bird is less, thus reducing per bird cost of housing. Cage systems are readily adapted to labor saving in materials handling. The use of automatic feeders and waterers, the belt collection of eggs and manure-handling devices are a natural with cages. Another important advantage of cage units is that eggs do not have to be collected as frequently because of the egg roll-a-way feature. Cage housing and management systems permit the care of more birds per man than does the floor system. For the most part we find that cage facilities offer better control of parasites and eliminate many of the troublesome litter-borne diseases that we sometimes find with floor birds. Most breeding farms still use the floor systems of housing for the mating of the breeders and the production of fertile eggs. Breeding cages are available, and may be successfully used for light breeders.

The requirements for a poultry house, or poultry pen, are that it provide a clean, dry, comfortable environment for the birds throughout the year. If young birds are to be brooded in the house, it must be built well enough to permit a comfortable environment and efficient use of fuel. It must be tight enough to prevent drafts on the birds and help maintain a uniform temperature economically. Poultry facilities should have floors which can be easily cleaned and disinfected. Most poultry pens or buildings should provide for ventilation to control moisture in the pens, remove gases and conserve or dissipate heat. In the northern climates the houses should have insulation in both the sidewalls and ceilings. Piped-in water is desirable but may create problems in severely cold climates, especially in the small flock situations where body heat is not sufficient to keep the pen temperatures above freezing. It is possible to correct this particular problem through the use of electric heat tapes. Houses should have artificial light to provide the right lighting program for both layers and growing stock. The poultry house should be strong enough to withstand high winds and snow-loads if necessary. The house plan (*Figure 1*) is for a small flock and provides room for storage of feed and supplies.

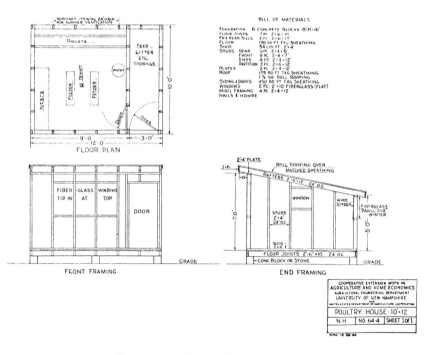

(Figure 1)

(Elevation of Figure 1)

Selecting A Site For The Poultry House

When selecting a site for a poultry house consider the soil drainage, air movement, location of the dwelling and the water supply. Remember there is the possibility of such problems as odors, flies, rats and mice. Good soil drainage assures dry floors which will help prevent wet litter, dirty eggs, disease, or other problems.

The poultry house should be located where the prevailing summer winds will not carry odors to the dwelling. The site should be large enough to provide for expansion, if this should seem advisable in the future. The location of access roads, electric lines, future buildings and yards should also be considered.

A site on relatively high ground with a south or southeast slope and good natural drainage is desirable. The location of a poultry house at the foot of a slope where soil or air drainage is poor, or where seepage occurs, is unwise.

If the house is located on a hillside the site should be graded so as to carry surface water away from the building. In some instances, tile drains will be necessary to adequately carry the water away from the foundation of the building.

In areas of severe winter weather, most of the windows should be in the front of the building, the house should face south to take advantage of the sunlight. The pen will be warmer, litter drier and the birds more comfortable. The closed or back side of the house should provide maximum protection against the northwest wind. Where the prevailing winter storms are from the west, however, the house should be oriented to face east. Wider houses are often placed with their long dimensions north and south so as to provide lighting on both sides. If the house is oriented with the front window area to the south, solar heat can be controlled by the width of roof overhang so as to provide shade in the summer months when the sun is high on the horizon, and permit sunlight in the pens in the winter when the sun is low.

space requirements

The size of the poultry house will depend upon the type and number of birds to be housed as well as the management system to be used. Various age groups and species of birds require different amounts of floor space for optimum results. If the house is designed for a laying house, and a cage management system is to be used, the size of the building will depend upon the dimensions of the cages, the type of cage, as well as the number of birds to be housed per cage. In addition to that space needed for the birds, there should always be some additional space for storage of feed, supplies and equipment.

The floor space requirements vary according to the type of project, the size or age of the birds and the management system in use (*Table 1*).

insulation and ventilation

The removal of moisture from the building is one of the main problems with keeping poultry. The moisture in fresh-voided manure is 70-75%. Then too, there is the respired moisture and moisture in the incoming air which must be taken care of.

Insulation will provide optimum bird comfort and avoid excessive moisture problems if the heat is adequate. In most areas of the United States the poultry house should be insulated. The purpose of insulation is to conserve heat in the cold climates and to keep out the heat in warm climates. It should be pointed out that light colored reflective roof and wall surfaces are helpful in keeping the pens cooler in warm climates. Since warm air carries more moisture than cold air, insulation serves to conserve heat in cool climates to not only provide bird warmth, but also to permit

TABLE I

FLOOR SPACE REQUIREMENTS FOR POULTRY

Type of Bird		*Floor Space (Square feet)
Chicks	0-10 weeks	.8 — 1.0
	10-maturity	1.5 — 2.0
Layers	Brown Egg	2.0 — 2.5
Layers	White Egg	1.5 — 2.0
Layers	Meat-Type Breeders	2.5 — 3.0
Broilers	0-8 weeks	.8 — 1.0
Roasters	0-8 weeks	.8 — 1.0
Roasters	8-12 weeks	1.0 — 2.0
Roasters	12-20 weeks	2.0 — 3.0
Turkeys	0-8 weeks	1.0 — 1.5
Turkeys	8-12 weeks	1.5 — 2.0
Turkeys	12-16 weeks	2.0 — 2.5
Turkeys	16-20 weeks	2.5 — 3.0
Turkeys	20-26 weeks	3.0 — 4.0
Turkeys	Breeders (Heavy)	6.0 — 8.0
Turkeys	Breeders (Light)	5.0 — 6.0
Ducks	0-7 weeks	.5 — 1.0
Ducks	7 weeks maturity	2.5
Ducks	Breeders (Confinement)	6.0
Ducks	Breeders (Yarded)	3.0
Geese	0-1 week	.5 — 1.0
Geese	1-2 weeks	1.0 — 1.5
Geese	2-4 weeks	1.5 — 2.0
Geese	Breeders (yarded)	5.0

*Many factors determine the floor space requirements including the type of management system, the type of house, the number and kind of bird, the climate and even the management the birds receive.

removal of moisture from the building. It is important to keep the litter dry and to prevent the buildup of ammonia in the house. Excessive ammonia can be detrimental to the birds and uncomfortable for the operator.

It was once thought that only housing in the northern climates need be insulated. The development of the so-called controlled environment house for high density operations has changed this philosophy, and we now find more of this type of housing being built in the south as well as in the northern climates, particularly for cage management systems. Controlled environment houses are windowless structures, with artificial light and fan ventilation.

Climate influences poultry house design, especially for the small flock. The farm building zone map (*Figure 2*) shows the four basic climatic zones in the United States, based on January temperatures and relative humidity.

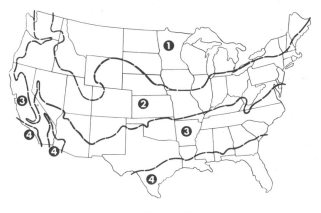

(Figure 2)

In the northern part of the United States the small flock,(less than a few hundred birds), is not without its special problems. Even the well insulated poultry house may not provide optimum conditions during severe winter weather unless artificially heated. It is essentially impossible or highly impracticable to insulate buildings well enough to conserve the heat given off by a small number of birds. Frozen waterers, frozen pipes or even frozen combs may result, causing a drop in egg production. Cold buildings also cause excessive feed consumption. Well insulated pens within large existing buildings may be one answer to some of the above mentioned problems, especially if other animals are housed in the building and help provide some of the warmth within the structure. Houses that are well protected from the wind are also much easier to keep warm and comfortable in winter weather. *Table 2* gives the recommended insulation values for poultry houses in the different climatic zones. The resistance or insulation values of various building materials is shown in *Table 3*.

TABLE 2

RECOMMENDED MINIMUM INSUALTION VALUES FOR POULTRY HOUSES BY ZONES

Location	R-value	
	Walls	**Ceilings**
Zone 1:		
Colder parts..................................	8-10	10-15
Warmer parts................................	6	12
Zone 2..	5	10
Zones 3 and 4....................................	2	5

Insulation value (R-value) is defined as the number of degrees difference in temperature between the inside and outside surfaces of a wall that will permit 1 British thermal unit (B.T.U.) of heat to pass through 1 square foot of the wall per hour.
Source: *USDA*

In zone 4, laying houses may be built with wire walls which can be covered with curtains during cool or windy weather. In zone 3 uninsulated houses with large openings in the front are frequently used. These are usually equipped with windows or plastic curtains for use in the cold or stormy weather.

TABLE 3

R-VALUES OR INSULATION VALUES OF
VARIOUS BUILDING MATERIALS

Material	Thickness in inches	Resistance or R-Value
Air space	3/4-4	.91
Asbestos cement board	1/4	.13
Blanket Insulation (Glass or Rock Wool)	1	3.70
Blanket Insulation (Glass or Rock Wool)	2-1/2	9.25
Blanket Insulation (Glass or Rock Wool)	3	11.10
Building Paper		Negligible
Cinder block	8	1.73
Concrete	10	.80
Concrete block	8	1.11
Fill Insulation		
Shavings	3-5/8	8.85
Shavings	5-5/8	13.70
Sawdust	3-5/8	8.85
Sawdust	5-5/8	13.70
Fluffy Glass Fiber, Mineral or Rock	3-5/8	13.40
Homosote	1/2	1.22
Insulation Board (Typical Fiber)	1/2	1.52
Insulation Board (Typical Fiber)	25/32	2.37
Plywood	3/8	.47
Sheathing and Flooring (Softwood)	3/4	.92
Shingles (Asphalt)		.15
Shingles (Wood)		.78
Siding (Drop)	3/4	.94
Siding (Lap)		.78
Surface (Inside)		.61
Surface (Outside)		.17
Window (Single Glass)		.10
Window (Double Glass in Single Frame)		1.44

ventilation

The purposes of ventilation are to provide bird comfort by removing moisture and ammonia from the building, to provide an exchange of air, and to control the environmental temperature in the pen.

There are two types of ventilation systems commonly used in poultry houses: the natural or gravity system and the forced-air system.

The gravity system may utilize windows, special slot inlets, or flues to provide air movement. The incoming cool air replaces the warmed, moisture laden air going up through flues in the ceiling or through the top of opened windows. If flues are used in cold climates, they must be insulated to prevent condensation of moisture which drips back into the pen. Flues must also extend well above the roof to provide the necessary draft to remove the warm, moisture laden air. This approach to ventilation works satisfactorily in an insulated house where inlets or windows are properly adjusted. The gravity method of ventilation has been gradually replaced as the size of poultry operations have increased. The need to open and close windows with changes in temperatures and wind conditions is a very time-consuming chore. Maintenance of the windows is also time-consuming and costly.

Forced-air ventilation gives the best control of air movement. Fans are used to move out the warm, moisture laden air and ammonia, and bring in cool air for bird comfort. Fan capacity is determined by the type, age, and number of birds in the house. For example, a house to be used for 2,000 growing or laying birds would be designed with a fan capacity of from 500 to 10,000 cubic feet per minute (cfm). This flexibility of air movement can be accomplished with either multiple fans, two-speed fans, or a combination of both. It permits a small amount of air movement for young chicks or for cold weather conditions, and a larger volume of air movement for older birds or warm weather conditions.

In cold climates 5 cfm would be too much air movement for bird comfort, so the ventilation rate is cut back to a lesser rate by the use of thermostats which control the fans. In cold climates the optimum temperature for all but young birds under 6-7 weeks of age is approximately 55° F. Fans should not be allowed to cycle on and off at frequent intervals. Fan cycling occurs when the pen temperatures are cooled down too quickly by over-ventilating. The maximum amount of moisture is not removed from the pens, and the birds are exposed to uncomfortable temperature conditions when this occurs. Fan cycling can be prevented with proper thermostat settings, and by restricting the air inlets in cold weather, to maintain a rather definite relationship between pen temperature, incoming air, and exhausted air. In the warm months the inlets are opened wide to permit a maximum flow of air for removal of moisture and ammonia and for cooling the birds. Houses for the small flock can usually be ventilated satisfactorily through the windows.

management systems

There are essentially three types of management systems used for the care of poultry. The one most widely used, until a few years ago, was the floor or litter management system. This is still the most popular method for small flocks, breeder flocks, growing birds, broilers, roasters, turkeys and waterfowl. This system permits the birds freedom of the entire pen or building. Litter materials such as wood shavings, ground corn cobs, sawdust, sugar cane or finely chopped straw are used on the floor. In addition to the necessary feeding and watering equipment, this system may utilize nests or roosts or dropping pits, depending upon the type of bird housed.

Another type of management system utilizes slats or wire mesh as the floor. This system has been used for laying birds, turkeys and waterfowl. The entire floor is slats or wire. No litter is used and roosts are not required. The bird's droppings fall through the wire or slats into a pit and are generally cleaned out periodically or when the birds are removed. A variation of this method incorporates either slats or wire and a litter section. The feeders and waterers are usually placed over the wire or slat sections, and the fecal material drops into a shallow or deep pit. The pit and the slatted or wire section are usually located in the center of the pen with a litter section on each side. This system has several advantages, in that the birds spend approximately 75% of their time on the slats or wire sections eating and drinking. Therefore, the majority of the droppings go into the pits, helping to keep the litter sections dry. This greatly simplifies the job of ventilating the building and results in cleaner litter and cleaner nests and eggs. The same types of feeders, waterers and nests are used for litter floor houses as for the slat or wire-floor houses.

The third type of management system utlizes cages. As mentioned earlier, cages have become the most popular means of managing market egg birds. Most of the large commercial laying houses are now equipped with wire cages. Cages have sloped floors so that eggs roll to the front for easy collection. Since the birds can't set on the eggs, they need not be collected as frequently. Cages are also coming into use to brood pullets to be housed in laying cages.

TABLE 4

APPROXIMATE SPACE REQUIREMENTS
(PER 100 BIRDS)

Type of Bird	Age (Weeks)	Brooder Hover-Type
Chicks (including broilers and pullets)	0-2	7 sq. in.
	2-6	7 sq. in.
	6-maturity	discontinue
Layers (all mash systems)	22-market	discontinue
(mash & scratch system)		discontinue
Light Roasters	0-2	7 sq. in.
	2-6	7 sq. in.
	6-maturity	discontinue
Heavy Roasters and Capons	0-2	7 sq. in.
	2-6	7 sq. in.
	6-13	discontinue
	13-maturity	discontinue
Turkeys	0-2	12 sq. in.
	2-4	12 sq. in.
	4-6	12 sq. in.
	6-16	discontinue
	16-market	discontinue
Ducks	0-2	12 sq. in.
	0-4	12 sq. in.
	4-	discontinue
Geese	0-2	12 sq. in.
	2-4	12 sq. in.

Infrared Lamps	Feeder	Waterer
2.............................	One feeder lid to 10 days....	20″ or 2 — 1 gal.
....	or...	
....	100″..	. . . .
2.............................	200″..	40″ or 3 — 1 gal.
discontinue..............	300″..	96″ or 1 automatic
discontinue..............	400″..	96″ or 1 automatic
discontinue..............	360″..	96″ or 1 automatic
2.............................	One feeder lid to 10 days....	20″ or 2 — 1 gal.
	or...	
	100″..	
2.............................	200″..	40″ or 3 — 1 gal.
discontinue..............	300″..	40″ or 3 — 1 gal.
2.............................	One feeder lid to 10 days....	20″ or 2 — 1 gal.
....	or...	
	100″..	
2.............................	200″..	40″ or 3 — 1 gal.
discontinue..............	300″..	40″ or 3 — 1 gal.
discontinue..............	400″..	96″ or 1 automatic
2-3 depending on..	One feeder lid to 10 days....	36″ or 4 — 1 gal.
outside temp.	or 192″....................................	
2-3 depending on..	288″..	72″ or 5 — 1 gal.
outside temp.		
2-3 depending on..	384″..	96″ or 1 automatic
outside temp.		
discontinue..............	480″..	120″ or 1 automatic
discontinue..............	600″..	120″ or 1 automatic
3-6 depending on..	100″..	20″ or 2 automatics
outside temp.		
3-6 depending on..	150″..	50″ or 4 automatics
outside temp.		
discontinue..............	....	50″ or 4 automatics

Furnishing The Poultry House

Furnishings for the poultry house depend upon the type and size of the project. If it is a growing type project the major requirements will be brooding equipment, feeders and waterers. A laying bird project will require feeders, waterers, nests, roosts and a broody coop for birds that go broody. If laying cages are used, then no nests, roosts or broody coops are needed.

The equipment need not be fancy. Feeders and waterers should be designed so as to service the birds efficiently with a minimum of waste or spillage. There must be enough equipment to give each bird an equal chance to utilize it. The amount of equipment required varies with the type of bird, age and size *(Table 4. See Preceding Pages).*

Equipment can be purchased from local feed and farm supply outlets or from mail order houses and farm equipment concerns (See Sources of Supplies and Equipment Section, page 211). For the small poultry operation much of the equipment may be homemade or purchased second-hand.

feeders

Feeders should be large enough to supply the flock's needs for a day or more without wasting feed. Proper feeder design is an important consideration in avoiding feed waste. The use of an anti-roost device, such as a reel or a spring-loaded wire on top of the feeders, will help to prevent waste. A lip on the side of the feed hopper will prevent birds from beaking out feed. Feed hoppers should never be filled more than 1/3 to 1/2 full to avoid feed waste. The size, height, and the construction of the feeder is very important if feed waste is to be kept to a minimum. To minimize feed waste, troughs should be made so that the height can be adjusted as the birds grow. The alternative is to change to a larger feeder as the birds become larger. Less feed is wasted if the lip of the feeder is level with the top of the bird's back. Some of the feeders commonly used and wooden feeders you can build are shown in *Figures 3* through *3e* on pages 22 and 23.

waterers

Birds of all ages should have access to plenty of clean, fresh water. Bear in mind that water is one of the cheapest sources of nutrients, and over one-half of the bird's body is water. Eggs are composed of about two-thirds water. Water is a very vital ingredient for all of the bird's body functions.

Water consumption depends upon the environmental temperature, age and species. Laying birds will drink about 2 pounds of water for every pound of feed consumed. In extremely hot weather water consumption may amount to 4 pounds for each pound of feed eaten.

For young chicks the waterers can be one-gallon glass jars with plastic bottoms. These are easy to clean. As the birds become older these should be exchanged for larger metal fountains or troughs. Older birds can be watered in open pans, pails or troughs. The containers may be equipped with a float to make them automatic.

If running water is piped to the house you may want to consider buying an automatic water fountain. The cost is usually not excessive, the amount of labor required is reduced and a constant supply of fresh water is available at all times. A homemade waterer (*Figure 4*) may be made from a gallon oil can. *Figures 4a* and *4b* show types of waterers commonly used for small flocks of poultry.

A fountain should be placed on a platform *(Figure 5)* or water stand, covered with a 1" x 2" mesh welded wire. The platform can be made of 2" x 4" material 30" to 36" square. For best results, it should be raised 3" to 4" higher than the depth of the litter. Better still, it can be placed over a drain if available. This arrangement will lessen he amount of wet litter surrounding the fountain and also help to keep litter material out of the fountain.

brooders

If day old birds are to be started, some type of brooder (heat source) will be needed. Enough brooder space must be provided so that every bird can get under the heat source as needed.

Brooders may be the hover-type heated by gas, oil, electricity, wood or coal. Infrared lamps work very well for small flocks and have the added advantage of enabling you to observe the chicks at all times.

FEEDERS

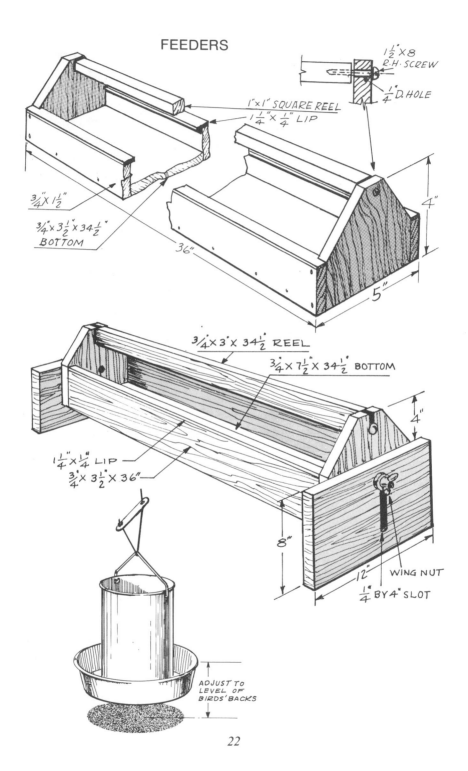

1½" X 8 R.H. SCREW

¼" D. HOLE

1" x 1" SQUARE REEL

1¼" X ¼" LIP

¾" X 1½"

¾" X 3½" X 34½" BOTTOM

36"

4"

5"

¾" X 3" X 34½" REEL

¾" X 7½" X 34½" BOTTOM

4"

1¼" X ¼" LIP

¾" X 3½" X 36"

8"

12"

WING NUT

¼" BY 4" SLOT

ADJUST TO LEVEL OF BIRDS' BACKS

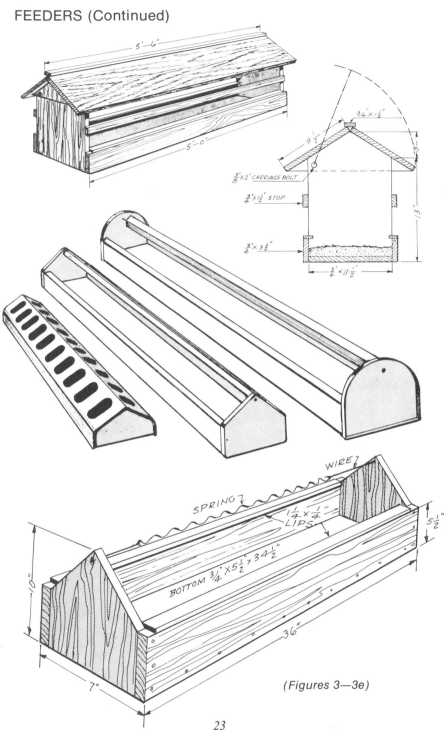

(Figures 3—3e)

WATERERS

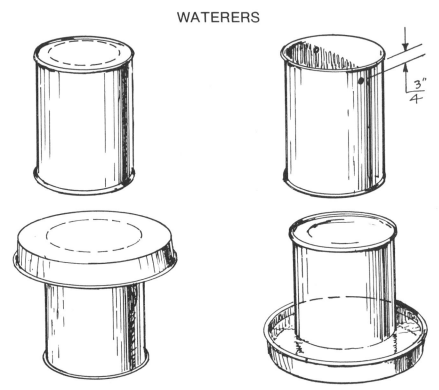

$\frac{3}{4}"$

(Figure 4. Homemade Waterer made from a gallon oil can and a pan.)

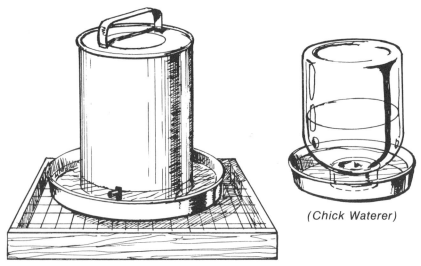

(Figure 4a. Waterer for older birds.)

(Chick Waterer)

WATERERS (continued)

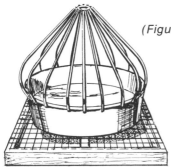

(Figure 4 b. Waterer suitable for laying birds.)

(Figure 5. Wire Water stand.)

   Figure 6 shows an infrared brooder with guard. *Figure 6a* is a plan for
a homemade electric brooder for a small flock. The typical hover-type
(Figure 6b) reminds one of a flying saucer. It is usually suspended from the
ceiling by chain or cable.

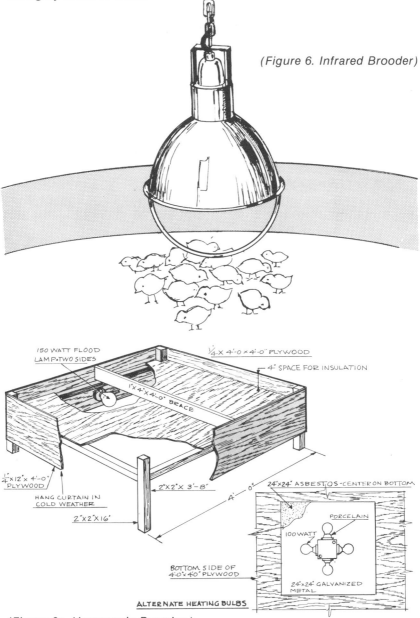

(Figure 6. Infrared Brooder)

(Figure 6a. Homemade Brooder.)

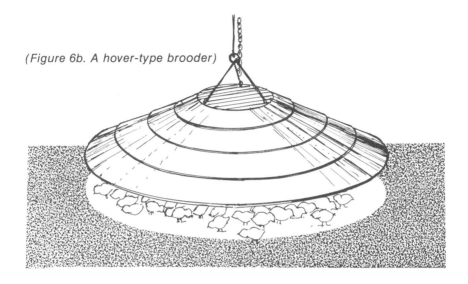

(Figure 6b. A hover-type brooder)

roosts

Roosts are not essential but are recommended for most small-laying flocks. Roosts should not be used for broilers or other meat birds. They may cause breast blisters. Roosts are used for turkeys but not waterfowl. Pullets should have 6"-8" of roost space per bird. If the layers are expected to roost they must be trained at an early age, or it will take considerable time to train them to use the roosts in the laying house.

There are several types of roosts. One of the more common types is the dropping pit with perches on the top of it (*Figure 7*). This type of roost is designed to accumulate droppings underneath for several months. This, naturally, keeps a large amount of the moisture out of the litter. If properly constructed and screened in around the top and sides, it will prevent the birds from getting into it. The nesting material and litter stays cleaner. Dropping pits are usually located either in the center of the building or at the rear of the building. These are normally the more comfortable areas in the house. The dropping pit should be designed so as to facilitate culling and so that it can be easily moved for cleaning. In larger units the dropping pits should be narrow enough to facilitate easy catching or culling of birds at night.

27

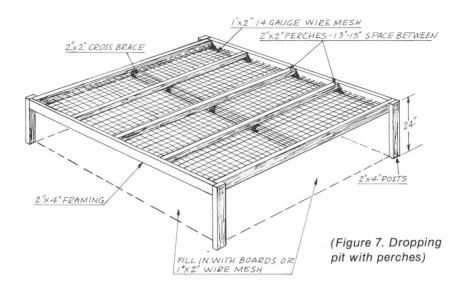

2"x2" CROSS BRACE

1"x2" 14 GAUGE WIRE MESH

2"x2" PERCHES - 13"-15" SPACE BETWEEN

24"

2"x4" POSTS

2"x4" FRAMING

FILL IN WITH BOARDS OR 1"X2" WIRE MESH

(Figure 7. Dropping pit with perches)

Another roosting method uses a dropping-board. The typical dropping-board is usually mounted on the wall at the heighth of about 2½ feet. The open space under the dropping-board provides additional floor space for the birds. The roosts are located above the dropping-board and the perches are hinged to the wall so they can be raised when the dropping-board is cleaned. This method is not as desirable as the dropping-pit type and is not recommended unless floor space is at a premium.

The roosts or perches should be of 2″ x 2″ stock, rounded or bevelled on the upper edges, so as to prevent injury to the breast and feet. For small breeds, allow 8″ of perch space per bird is needed. Large breeds should have 8 to 10″ of perch space. Roost perches are usually spaced 12″-15″ apart.

nests

An adequate number of well designed nests with clean nesting material should be provided for the laying hens. They can be either individual or community-type nests. One individual nest, or one square foot of community nest, should be provided for each four laying birds. Individual nests (*Figure 8*) should be at least one-foot square and one foot high. Community nests (*Figure 8a*) can be almost any size but must be provided with at least two openings, 9″ x 12″ for every 20 square feet of nest space.

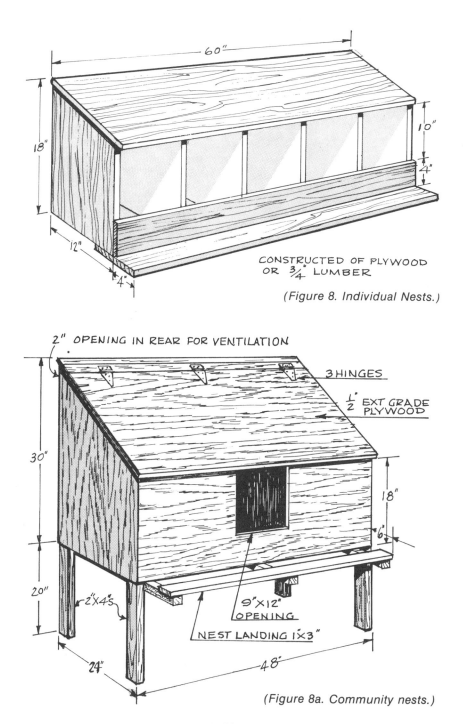

60"

18"

10"

4"

12"

4"

CONSTRUCTED OF PLYWOOD
OR 3/4" LUMBER

(Figure 8. Individual Nests.)

2" OPENING IN REAR FOR VENTILATION

3 HINGES

1/2" EXT GRADE
PLYWOOD

30"

18"

6"

20"

2"X4'S

9"X12"
OPENING

NEST LANDING 1"X3"

24"

48"

(Figure 8a. Community nests.)

As an aid to getting the birds into the nests, they can be darkened by covering two-thirds of each nest entrance with a cloth flap. A landing board should be placed below the nest openings to provide easy access to the nests. The nests should be located approximately 2 feet from the floor or surface of the litter.

Commercially-made nests are also available from agricultural supply houses and specialized equipment dealers. Some of these people offer a nest, either a community or individual type, with a roll-a-way feature. After the eggs are laid, they roll out to the front or back of the nest into an egg tray where they can be easily gathered. A word of caution about this type of nest: it's excellent in theory, but birds must be trained to use it and, more often than not, a large percentage of the eggs are laid on the floor rather than in the nest.

Nests should be cleaned frequently, as the nest material becomes dirty. Since several hens use the community-type nest at one time, ventilation of the nest is very important. For this reason a 2″ wide opening at the top of the nest is provided to allow heat to escape.

cages

Laying birds may also be housed in wire cages. Cages may be located in an open building or in closed fan-ventilated housing. As mentioned earlier, cages simplify the management of layers, but birds in cages are more susceptible to the effects of extreme weather conditions. Therefore, they must be protected from wind, cold and, particularly, from hot weather.

Cages (*Figure 9*) are usually constructed of 1″ x 2″ welded wire, the floor is constructed with a slope of approximately 2 inches per foot, so that eggs will roll out onto the egg tray. Feed and water is provided in troughs usually located on the outside of the cage. Water cups located inside the cage are also sometimes used.

There are several types of cages available, most of which are quite satisfactory. They come in several sizes but should not exceed 20 inches in depth for best results. Cages 12 inches wide and 18 inches deep are very popular. Research has indicated that three brown-egg birds or four Leghorn-type birds can be placed in a 12 x 18 inch cage with good results.

Several types of cages are satisfactory. They include the stair-step, the double-deck with dropping boards, colony and single-deck. Stair-step cages are available in two arrangements, either full or modified. Some commercial producers are also moving toward a triple deck or even a four-deck modified stairstep-cage system to enable them to house more birds in a given area.

The use of cages cannot be justified in many of the very small laying flock enterprises. However, there are situations in which they can be used to good advantage. We have helped plan several small cage units for flocks of three or four hundred birds, utilizing an existing building or pen space more efficiently than could have been accomplished using a floor system. In these small units the birds are usually watered automatically, but the feeding, egg collection and cleaning are manual chores.

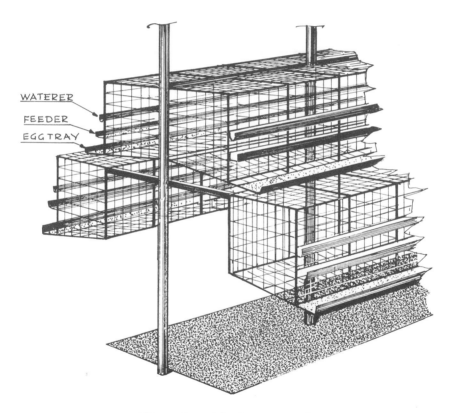

WATERER

FEEDER

EGG TRAY

(Figure 9. Full Stair-step cage)

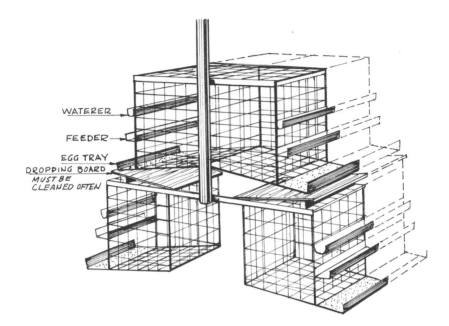

WATERER

FEEDER

EGG TRAY
DROPPING BOARD
MUST BE
CLEANED OFTEN

(Figure 9a. Modified Stair-step cage)

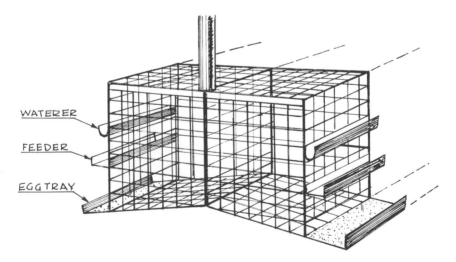

WATERER

FEEDER

EGG TRAY

(Figure 9b. Single-deck cage)

3

STARTING
THE LAYING FLOCK

The laying flock can be started in a variety of ways. You can set eggs under a broody hen or in a small incubator. Either day-old chicks or started pullets may be purchased. One questionable method of starting is to purchase second-year layers. These are birds that have laid for a year. Production is not as good for second year layers, egg quality may be poor and they may stop laying, go into a molt (shed old feathers and grow new ones) soon after moving. One of the simplest and best ways of starting the laying flock is to buy started pullets. This avoids the need for equipment and the care involved in incubating eggs and brooding chicks. You can buy started pullets from commercial growers at 19 to 20 weeks of age. Sometimes it is possible to buy a few surplus pullets from a local poultryman in your area. Pullets should be debeaked to avoid feather picking and cannibalism. Ten to 15 birds will normally provide the average family with enough eggs. Unless surplus eggs can be marketed easily you should not produce more than the family needs.

Before buying birds, decide whether you want white or brown eggs. Egg shell color does not affect food value, but it does influence the market price, since there is a definite preference in various areas of the country. In most sections of the United States, white shell eggs sell for slightly more than the brown shell eggs. In the New England area, or so-called Boston market area, brown eggs are preferred and the situation is reversed.

Birds that lay white eggs usually weigh in the vicinity of 4 to 4½ pounds at maturity. Representatives of these breeds and varieties are the Leghorns, first generation Leghorn strain crosses or hybrids — all of which are good egg producers but not so good for meat.

The medium weight breeds, American Breeds, lay brown eggs. The mature weight of these birds is usually between 5 and 6 pounds, thus making them more desirable as meat birds. Most of the medium weight, brown egg producing layers are crosses of American breeds, such as the Barred Plymouth Rock and the Rhode Island Red. There are several of these crosses, all of which are good egg producers. There are several pure breeds such as the Rhode Island Reds, White Plymouth Rocks and New Hampshires, which can be used as dual purpose breeds, that is for meat and eggs. However, they may not do as well as the crosses which have been bred especially for meat or egg production.

Stock that is bred for high egg production or for meat production is best suited for the home flock. To get the project off to a good start, purchase well bred healthy chicks or started pullets. One of the mistakes that is frequently made is the purchase of chicks from distant hatcheries. Sometimes the conditions to which chicks are exposed when being transported over long distances are less than desirable. They may be chilled or overheated, leading to early losses and a poor producing flock. It is usually best to purchase chicks or started pullets from a reputable nearby hatchery or producer that has stock bred for efficient production. Traits that should be considered, when purchasing chicks or pullets, include liveability, early feathering and feed efficiency, freedom from disease, rate of growth, egg production and egg quality. Much valuable information can often be obtained by talking with poultry producers in your area. They can frequently tell you from experience which breeds and strains have the above characteristics and the most reputable hatcheries and pullet growers from which stock can be obtained. Another means of evaluating stock is to consult the random sample egg production and meat production test reports. Random sample egg and meat production tests are conducted in several states. These tests are designed to test the entries of production stocks for the various economic production traits. The test reports provide information to help evaluate the performance of laying stock or meat stock offered for sale by the participating breeders and hatcherymen.

The United States Department of Agriculture annually publishes a report on these production tests. Copies of the publication, which contains records of all stocks entered in the performance tests in the United States and Canada, may be obtained by writing to the Poultry Research Branch, Animal Sciences Research Division, Agricultural Research Service, United States Department of Agriculture, Beltsville, Maryland.

One of the first steps toward getting disease-free chicks is to buy the stock from hatcheries that blood test their breeders for pullorum-typhoid. Hatcheries which operate under the National Poultry Improvement Plan do blood test their breeding flocks for these diseases. Both of these diseases are passed from the infected hen to the chick through the egg.

If a nearby hatchery has stock with good production capabilities and is disease free, it is best to get your chicks locally to minimize the shipping time and avoid the possibility of prolonged exposure to extremes of temperature and poor handling. It is usually much easier to get adjustments from a nearby hatcheryman in the event that problems arise. Chicks should be ordered at least four weeks in advance of the date you would like to start them, and started pullets should probably be ordered close to six months before they are needed. Your County Extension Agent or Extension Poultry Specialist can help you select sources of stock. (See page 203 for Directory of State Poultry Specialists.)

Buying Your Chicks

The importance of selecting the right stock should be re-emphasized here. After deciding on the breed or strain you want, be sure to buy healthy stock from hatcheries which have a U.S. pullorum-typhoid clean status. Buy the best chicks you can buy. Cheap chicks may cost you more in the end than those that cost most in the beginning. Poor results with growth or liveability can soon erase any savings that you may have made in the purchase of chicks.

Be sure to buy chicks which are bred for the purpose for which they are to be used. If you are interested in starting with a flock of laying birds, get chicks from a source which is known for having birds with high egg production records. There are great differences among types of birds, insofar as egg size, egg production, feed efficiency, liveability and many other economic traits are concerned.

In the New England area, production birds that lay brown eggs are primarily the sex-link crosses. There are about three or four birds of this type offered for sale in New England. They are known as sex-link crosses, because the breeds that are mated result in male chicks of one color and female chicks of another color. These can be sexed by feather color at day-of-age, thus eliminating the cost of vent sexing. Some of the breeders of inbred crosses, noted mainly for their white-egg birds, also have inbred crosses which produce brown eggs.

When buying chicks for laying stock the usual practice is to purchase sexed pullets. This is because the cockerels of the best laying stocks are apt to be poor meat producers. It is usually not wise to try to grow them for meat.

If you are planning to raise meat-type birds, it is recommended that you order the birds which are bred for the production of meat. Only birds bred for meat production should be raised as broilers, capons or roasters.

Stock that is bred for the production of meat has the ability to grow rapidly with a minimum of feed consumption. Feed consumption represents at least 60% of the total costs of producing poultry meat. Thus feed cost becomes an important consideration. Also, birds that are bred for meat tend to mature with a better finish, are tenderer and more juicy. Then too, the meat bird stocks are primarily white feathered, which results in a much better dressed appearance in the absence of dark pinfeathers.

when to start the chicks

The time to start chicks may depend upon the type of housing you have and the type of brooding equipment available. Chicks started in the winter commence laying in the early summer, but may go into a neck molt and take a vacation the following winter. On the other hand, pullets, which are hatched in late winter or early spring, start to lay in the summer when egg prices are normally highest. Those raised late in the spring will not start laying until late fall when egg prices are usually lower.

Probably the best time for most small flock producers to buy their chicks is in late March, April and May, especially those in the northern climates. Chicks started at that time do not need cold weather brooding facilities. Moreover, the pullets will start laying in early fall and will continue laying for a full 12-month cycle.

4

BROODING AND REARING YOUNG STOCK

The Brooder House

Brooding facilities can affect the degree of success one has in growing chicks. As indicated earlier, an expensive building is not necessary, but the facility should be built so that it will brood chicks efficiently.

One of the features of a good brooder house is structural soundness to withstand high winds and snowloads in winter areas. It should be of tight construction, with sufficient insulation against drafts, and easy to maintain a uniform temperature economically. It should have a ventilation system to control moisture and gases, yet still able to conserve or dissipate heat. Lastly, it is desirable to have some degree of light control in the house. The main equipment needs include the feeders, waterers and brooders. Roosts may be used for replacement pullets.

Equipment

feeders

Commercial or homemade feed hoppers are satisfactory for use in brooding chicks. Probably the commercial hoppers are easier to clean and cause less feed waste.

waterers

It is possible to make water fountains using oil cans, or juice cans in combination with tin plates. This can be done by making two holes on opposite sides of the can about ¾″ from the lip. The can is then filled with water the plate placed on top, the can and plate is then flipped over to provide a waterer that maintains its own water level. See *Figure 4* for details on making this waterer.

A small wire platform underneath each fountain helps to keep the water clean, keep the shavings out of it and also prevents the litter from becoming too damp around it.

Probably the best watering devices are the commercial one-gallon fountains. Commercial poultry producers frequently use the gallon fountains in combination with automatic waterers. They gradually move the gallon fountains close to the automatics. They are thus able to convert to the automatic waterers as early as 10 days of age. Once the chicks are drinking well from the automatic waterers, the fountains can be removed gradually and additional automatic fountains added until they have three to four eight-foot waterers per 1,000 birds.

In small flock situations, two one-gallon waterers are usually good for 100 birds for up to 2 weeks. From the second through the sixth week, there should be at least two 2 or 3-gallon waterers, and from 8 to 10 weeks, one 5-gallon waterer per 100 birds.

brooders

To provide the necessary heat some type of brooding device is required. Infrared lamps are an excellent source of heat for brooding small flocks of chicks *(Figure 6)*. The initial cost of equipment is relatively small, and the lamps are excellent for supplying heat to the chicks, particularly in the late spring, summer, or early fall months. The infrared lamp is suspended approximately 18″ above the litter. One 250 watt bulb is sufficient to brood up to 75 chicks. Even though the number of chicks may be small, it is wise to use at least 2 bulbs in case one fails.

Many small flocks are brooded with homemade brooders utilizing electric light bulbs as the heat source. It is possible to construct simple hovers (canopies) of plywood and canvas, or some similar material, which can be suspended from the ceiling or supported by legs *(Figure 6a)*.

Other brooders which are manufactured commercially are fired by oil, gas, electricity, wood or coal. These have a hover to keep the heat down at the floor level *(Figure 6b)*. Such brooders are excellent for flocks of 100 chicks or more. Some of them are rated at a capacity in excess of 750 chicks. The initial cost of these units is considerably more than either the infrared lamp or the homemade brooder. Each chick needs a minimum of 7 square inches of hover space or its equivalent.

Each type of brooder has certain advantages. The infrared bulb has the advantage of enabling the operator to see the chicks at all times. Where the hover-type brooder is used, it is more difficult to observe the chicks. On the other hand, the hover-type brooder is best for cold weather brooding and is less expensive to operate.

The type of equipment that is selected will depend, to a certain extent, upon the cost and availability of fuel and the extent to which it is to be used. Whatever brooder stove you use, it should have the capacity to heat adequately and a means for controlling the temperature accurately.

roosts

Roosts should not be used for broilers, roasters or capons. They can cause breast blisters or breast deformities. In fact, few people use roosts even for growing replacements. Those who want their birds to roost in the laying house feel that it is necessary to train the pullets to roost in the brooder house. If roost perches are to be used in the brooder house, they should be enclosed with poultry netting so as to prevent the chicks from coming in contact with the droppings. They should be constructed so as to provide easy access for the birds. Frequently, they are slanted from the floor to the wall much like a ladder.

5

BROODING MANAGEMENT

Preparation Of The Brooding Quarters

G etting the chicks off to a good start is very important. Probably no other part of the enterprise deserves more planning and advance preparation than does the starting of baby chicks. The modern brooder house should be clean and disinfected or fumigated at least two weeks before the chicks arrive. If this is not done, the chicks may be exposed to certain poultry diseases which can lead to mortality and poor results. First sweep or wash down cobwebs, dirt and dust, then use a good disinfectant. There are a number of good commercial preparations available. They should be used according to the manufacturers recommendations. Some of them require that the house stand idle for a period of time and be thoroughly aired before the chicks arrive. Failure to observe these precautions may cause the chicks to have severely burned feet and eyes.

After the brooder house has been cleaned and disinfected, it should be allowed to dry thoroughly before putting in new litter. When the floor is dry, it should be covered with 2"-4" inches of litter material. The litter material serves to absorb moisture and to insulate the floor for comfort of the birds. Several different materials may be used for litter. Wood shavings are among the most commonly used, whenever available. Other materials include sawdust, or shavings and sawdust, rice hulls, sugar cane, peanut hulls and ground corn cobs.

When chicks are extremely hungry upon arrival they may try to eat the litter before learning to eat feed. If you have reason to believe that the chicks have been hatched for two or three days when they arrive, it may be a good idea to put paper over the litter for the first 3 or 4 days until they learn to eat. When chicks are received soon after hatching, it is not necessary to do this.

The brooders should be started a day or two before the chicks are due to arrive. This will assure that the equipment is operating properly and is adjusted to the correct temperature. A chick guard should be used to confine the chicks to the source of heat. It will also help to prevent drafts on the chicks under the brooder. A corrugated cardboard guard, approximately 12" high is good for this purpose during cool weather. For warm weather brooding the guard may be made of poultry netting. It should form a circle around, and about 3 feet from, the brooder. The guard is usually removed after 10 days or so, depending upon the weather and conditions within the house. The chick guard keeps the chicks confined to a given brooder and also prevents migration and over-crowding of some brooders if more than one brooder is used in the brooding facility.

Fill the feeders and waterers several hours before the chicks arrive. The water will be at room temperature when the chicks arrive, and encourage them to drink. The feed trays and water fountains should be spaced uniformly around the brooder and close to the hover (*Figure 10*). One method that is frequently used to get chicks to eat, when first put under the hover, is to place a small amount of feed on either new egg filler flats or on inverted chick box lids. The type of lid that is used is the one without the slots or holes for ventilation. Chicks instinctively peck at anything at the same level as the surface upon which they are standing. Therefore, they will learn to eat more readily if feed is provided in this manner for a day or two. Chicks that are slow to catch on to the fact that feed is in front of them are attracted by the noise made by the others that have found the feed and are pecking on the cardboard. This gets the chicks off to a good start and eating well.

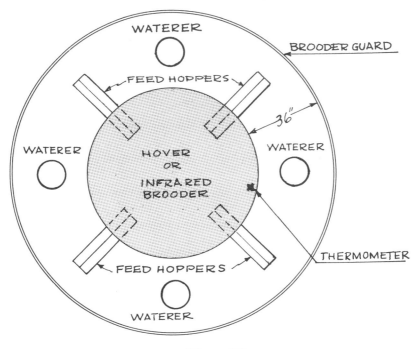

(Figure 10)

brooder management

Brooding temperature is very important. The recommended temperature at the start is 90-95° F. A rule of thumb to remember is: reduce the temperature 5° each week until the chicks no longer need heat. It is often stated that the best thermometer, to gauge the most comfortable temperature, is to watch the chick itself *(Figure 11)*. After the chick

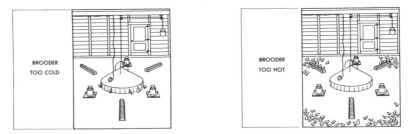

(Figure 11. Brooder Management)

42

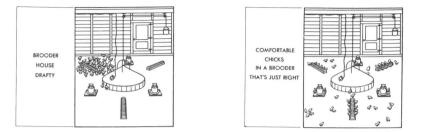

reaches 7 days of age this is probably true. If the chicks huddle close to the heat source you can be reasonably sure that the operating temperature is too low. If, on the other hand, the birds are located in a circle, way outside the heat source, you should assume that the temperature is too high. During the day the chicks should be evenly distributed around the entire brooding area, with some of the chicks underneath the heat source. Temperature readings should be taken at the outside edge of the brooder at the chick level. One important factor for the successful brooding of chicks is the maintenance of good litter conditions. When litter conditions get out of hand and become too wet, disease problems can result. Wet, dirty litter can harbor many disease organisms which affect poultry. In the case of replacement pullets, it is desirable to have a certain amount of moisture in the litter (30-35%) to enable sporulation of oocysts and the subsequent development of an immunity to coccidiosis during the growing period. However, an excessively wet litter, especially during warm periods, can bring about a clinical case of coccidiosis and result in what we term a coccidiosis break, which requires treatment.

One of the common problems experienced with meat birds is breast blisters — swellings or external sores on the skin of the breast — which seriously detract from the dressed appearance of the carcass. These are quite common in flocks of capons and roasters. Excessively wet, caked or dirty litter is frequently blamed for breast blisters. Since perches are not used for broilers or meat birds, the birds must bed down in the litter. It is important that the litter depth be maintained and that it is kept dry and fluffy enough to cushion the body and avoid all contact with the floor. Furthermore, it is essential that it be reasonably clean to avoid soilage of the breast feathers, skin irritations and breast blisters.

General Feeding Programs

The recommended modern feeding programs for meat birds and for laying bird replacements vary widely. Broilers and other meat birds are fed a high energy diet throughout the growing period. In the production of meat birds, the goals are rapid growth, heavy weights and efficient feed conversion. Replacement pullets, on the other hand, are started on medium-to-high-energy starter rations for the first 6 to 8 weeks; then they are changed to a grower diet to accommodate their changing dietary needs. Meat bird chicks and replacement pullets which are reared on litter are normally started on a diet containing a coccidiostat. A higher level of coccidiostat is used for meat birds to prevent coccidiosis, a disease common to chickens reared on the floor. The coccidiostat is usually fed throughout the growing period in the case of the meat birds to protect them against a coccidiosis outbreak and to obtain maximum growth and efficiency. Replacement pullets are fed a lower level of coccidiostat in the starter diet. This enables them to develop an immunity to coccidiosis by 6 to 8 weeks of age, and this immunity protects them throughout the remainder of the growing period and the laying period.

Diets for meat birds, particularly broilers and turkeys, are usually somewhat better fortified with vitamins and growth-promoting factors. They sometimes contain antibiotics. The protein level of meat bird diets, at least during the first several weeks, is higher than for replacement birds. Another difference is that feeding programs for meat birds do not include scratch grains during the growing period. Occasionally, heavy meat birds are fed corn the last two weeks of the growing period to improve their finish just prior to dressing or marketing. Frequently a special finishing diet is used for this purpose. Some growers of meat birds prefer the crumbles or pellets to the mash form because of less feed waste and increased feed consumption. However, crumbles and pellets cost more, and there is also the tendency for birds fed on crumbles, and particularly pellets, to feather-pick or become cannibalistic.

Some broiler or roaster feeds contain feed additives that must be removed several days prior to slaughter. The withdrawal period varies with the type of drug. For the last several days the birds are fed on a special feed without the additives. Some drugs used for the treatment of diseases are administered through the feed. Definite withdrawal periods are required for these drugs prior to slaughter. Always follow the recommendations of the feed company printed on the feed tag.

feeding and management of layer replacements

Replacement chicks for layers are started on a diet very similar to the broiler diet; however, the energy and protein content are usually slightly lower. The protein is 20 to 21%. Chicks to be raised for layers are normally fed a starter for 6-8 weeks and then changed to a grower diet.

As with other animals, the growth rate diminishes as the birds become older. Therefore, the protein requirement is less later in the growing period and the grower diet is formulated with somewhat less protein than the starter.

Several feeding programs are used from the 6-8 week period to housing. However, the two basic systems are the all-mash diet with approximately 15% protein or a mash and scratch diet. For a number of years it was felt that it was best to full-feed growing replacements. Today, more and more growers are using some form of feed restriction to delay sexual maturity, yet permitting the development of a healthy pullet. Other growers are using lighting programs to delay sexual maturity. No doubt the simplest and best procedure for the small flock manager is to full-feed the growing pullets. When feed is restricted, there must be adequate equipment and good management to avoid problems.

why delay sexual maturity?

Birds that commence to lay too early or that are too fat at the onset of production frequently do not perform as well as slower-maturing birds. They are inclined to lay fewer and smaller eggs and are more prone to prolapse of the uterus and higher mortality.

Birds on a full-feed program that are grown in the off-season (under conditions of increasing day-length) should receive a controlled lighting schedule to delay sexual maturity. This method, if done correctly, usually produces the most economical gains and more uniform birds with fewer social problems or vices.

There are several restricted feeding programs. One of the fairly common ones is the use of a 15 or 16% protein diet of medium energy. The exact diet depends upon the type of bird. Usually some physical restriction is used. It can be a skip-a-day program, or a specific amount of feed can be fed daily depending upon the weight and age of birds. The restriction requires attention to certain details. For example, it requires about 50% more feed space for a restricted feeding program to assure that all birds get feed. Then too, restricted programs tend to increase feather pulling and cannibalism and may increase the nervousness of the flock.

A feeding program which is still used, to a considerable extent, is a combination of the grower concentrate with a grain such as oats. After 8 weeks the oats can be gradually added to the grower diet until proportions of approximately one-half grower concentrate and one-half oats is reached. Some producers permit the birds to free-choice feed the mash and oats. Grit should be used when a whole grain is fed. A hard insoluble grit, not a soluble grit such as calcite crystals, should be used. Use a medium or pullet size grit to approximately 12 weeks of age and the hen size, or coarse, for the remainder of the growing period. Grit must be present in the gizzard to enable birds to grind the fibrous material so that it can be utilized and absorbed in the intestines.

Other restriction programs now receiving some attention for delaying sexual maturity in pullets, are high-energy, low protein-type diets and low lysine diets. Good pullets can be grown on most of these programs. Check with your feed company, and the breeder of the stock which you purchase, to determine which system he recommends.

Heavy breeds have a tendency to acquire more fat than the lighter breeds and crosses of egg-producers. These birds, and heavy broiler breeders, for hatching egg production are commonly put on a restricted program, usually an all-mash diet.

Range-Rearing

Under some conditions range-rearing or yarding can still be recommended. It may be possible to save some on the cost of feed and still grow good healthy pullets. Practically all broilers and most meat birds are reared in confinement, so as to have absolute control of the feed intake, more rapid and efficient growth and better finish. As mentioned earlier, rapid growth and early maturity, are not desired in replacement birds, so range-rearing or yarding may be used for replacement pullets.

Range-rearing was once a very popular method of growing young stock during warm weather. There is no question that excellent pullets can be grown on range. Good range-reared birds usually have excellent shank color, beautiful feathering, and are inclined to be big-framed birds having health and vigor. Why, then, have many producers gone to confinement-rearing?

Losses from predators, the increased labor requirements in feeding and caring for the birds, are probably two of the biggest reasons for getting away from range-rearing of pullets. Predators such as foxes, coons and even hawks and owls can be a real problem where birds are kept on range. Then there is the expense of building fences and range shelters and their upkeep. If the birds are to be yarded you can perhaps utilize the brooder house for shelter, thus avoiding the need for special range shelters. The choice of whether to range or confine birds will depend upon a number of factors, such as the size of flock, the labor situation, and the land and housing facilities available. *Figure 12* shows a range shelter suitable for about 100 pullets.

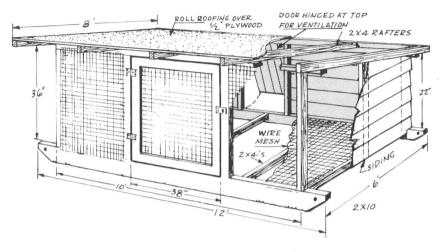

(Figure 12. Range Shelter)

A good range properly managed will yield a good supply of forage for growing pullets. Some of the grasses and legumes that are most satisfactory for poultry range include ladino clover, Kentucky blue grass and brome grass. This type of range permits feeding of a grower diet on a restricted basis and makes the birds forage for a portion of their nutrient intake. The range should be kept clipped to be most palatable and kept in a sanitary condition that is, prevent bare spots and mud holes.

Range shelters and equipment should be moved occasionally when the ground around them becomes bare, so as to keep the birds out of the mud. It is desirable to change range areas so that a given area is not used every year. This may help to avoid disease problems. Capillaria worms are one of the problems experienced on wet muddy range.

Covered feeders are recommended when birds are fed on range. They will help prevent feed loss from wind and rain. A covered range feeder is shown on page 23.

It requires an acre of good range to properly handle 400-500 pullets. Range-rearing is limited to the warm seasons of the year or to the areas which have warm climates.

Small flocks of layers, pullets, roasters or capons are frequently provided a yard. The yard is a fenced-in area attached to the brooder house or some other shelter. A yard permits more birds to be kept in a given floor area and may help prevent management problems such as cannibalism or feather picking. Its use in the cold climates is pretty much restricted to the warm months.

Where yards are used, the feeding and watering equipment is usually left in the house. If covered feeders are used some of the feeding may be done in the yard.

The same hazards may exist with yards as with range or pasture. Disease may be a consideration where the same yard is used year after year. Ideally the yard area should be moved each year. Predators may also be a problem. Layers which are permitted to yard may hide their eggs in areas outside the house. In wet weather the eggs may be excessively dirty. It is usually best to confine the laying flock for optimum results.

Vaccination

Several vaccines are available to prevent outbreaks of some of the troublesome poultry diseases. Some of the common ones available for immunizing birds are for infectious bronchitis, Newcastle disease, laryngotracheitis, fowlpox and cholera. In some areas, all of these diseases are a problem. Newcastle and bronchitis can be problems anywhere, and fowlpox is prevalent in many areas. In northern New England we don't see fowlpox very often and do not recommend vaccination for it. The same has been true for laryngotracheitis. Vaccination programs should be established on the basis of those diseases which are prevalent in the particular area. Commercial producers routinely vaccinate for Newcastle and bronchitis in all areas of the United States and, in some areas, also vaccinate for fowlpox and cholera on a routine basis. Vaccination recommendations in various areas, therefore, will differ, and the individual program should be developed to handle the disease situation in the particular area.

Some of these vaccinations come early in the chick's life, the Marek's vaccination, for example, administered at day of age, usually at the hatchery. Incidentally, it is advisable to purchase chicks that have been vaccinated for Marek's disease. Marek's disease did, for many years, take a tremendous toll on growing birds. With the advent of the Marek's vaccine, mortality has been cut to a very negligible rate. Other vaccinations are administered at various times during the growing period and may require booster shots during the laying period.

There are two basic types of vaccines, those which produce a temporary immunity and those which produce a solid or permanent immunity which will protect the birds for the remainder of their lives. The virus vaccines producing a temporary immunity are usually of low virulence, or, in some cases, have no apparent affect on the birds when administered. The virus of vaccines producing a permanent immunity are of a more severe virulence and therefore, will produce more marked or serious symptoms of sickness.

There are two types of vaccines on the market with regard to method of application. There are those that are applied to individual birds and those that are applied to the flock on a mass basis. Two of those that are applied on a mass basis are vaccines for Newcastle and infectious bronchitis. These may be applied as a spray, dust, or in the drinking water. Those vaccines that are used on a mass basis are mild vaccines of low virulence.

Contact your Poultry Diagnostic Laboratory or Poultry Specialist at your State University. Local industry service people and County Extension Agents can also be helpful. Some vaccines are not permitted in certain states. Vaccination of small flocks is not recommended in many areas.

Because there are so many factors to consider, and so many variable conditions from farm to farm, no set vaccination program can be recommended here that will best suit all farms. The program should be planned to meet the needs of the given area. It should be planned well in advance of the brooding and rearing season, and be scheduled so that all the birds that are to be placed in laying houses will have been vaccinated well before the time they are ready to lay. This will give them sufficient time to fully recover from the stress of vaccination and avoid any carry-over effects there might be on egg production or egg quality.

Feather Picking And Cannibalism

Both feather picking and cannibalism are rather common vices which can develop in the brooder house and may be carried on into the laying house. A similar type of problem is toe picking which can start in a flock of chicks soon after they are put down under the brooder.

The exact cause of these vices is not always evident. Some factors that are thought to contribute to them are poor nutrition, overcrowding, or other faulty management practices. Overheating the birds, lack of floor, feed and water space, or even overlighting, are all thought to be factors.

Sometimes the problem will occur even with apparently good management.

At the first sign of feather picking, action should be taken to stop it immediately. When blood is started the birds frequently pick each other apart and losses can be severe. In some flocks, the problem can be corrected by a change in management — providing more feed, water, or floor space, better ventilation, cutting back on the light, or any number of things that will improve the birds comfort or change environmental conditions.

In the case of small flocks, anti-pick ointments applied to the birds have been somewhat successful in curtailing the problem. Plastic and metal devices known as "specs" can also be used. "Specs" are mounted on the top beak and are designed to obstruct the bird's forward vision. Sometimes these devices catch on the poultry furniture and come off or partially unattach. The most widely-used, and probably the most satisfactory method of cannibalism control, is to debeak the birds. Many producers debeak their birds routinely to prevent picking problems.

Chicks can be debeaked at day old at the hatchery, using a precision debeaking machine. The beaks tend to grow out again. To be safe, most producers debeak again before the birds are put into the laying house. Unless the job is done properly, debeaking at day of age can result in chick mortality due to starve-outs.

A number of producers are now debeaking at 7 to 9 days of age. Special adaptors for debeaking machines make this a precision job. If done properly, this usually holds and the job need not be done again. Debeaking can be done at any age to correct a picking problem. However, it is best done before the birds start to lay and should be done by 16 weeks of age.

Debeaking involves cutting off slightly more than one-half of the upper beak and blunting the lower beak. The upper beak should be shorter than the lower beak, making it difficult for the bird to grasp feathers or skin. A temporary debeaking job can be done in the absence of a debeaking machine. This is done by the removal of a small portion of the upper beak with nippers or dikes. The beak soon grows out and the procedure needs to be repeated.

A debeaking machine has an electrically heated blade that cuts and cauterizes at the same time to avoid bleeding. Properly debeaked growing and adult birds are shown in *Figures 13* and *13a*.

(Figure 13. Properly Debeaked Pullet)

(Figure 13a. Properly Debeaked Adult Bird)

Lighting Programs For Young Stock

The first three days to a week, the birds should be provided with 24 hours of light. During this time they get used to their surroundings and learn where the heat, water and feed are.

After this time the chicks should be put on a controlled lighting program unless hatched during the April to August period. All pullet chicks grown in windowed houses and hatched between April and August can be exposed to natural daylight, because daylight length will be decreasing during the latter part of the growing period, thus delaying maturity. The rule of thumb to follow is never expose replacements to an increasing light day during the growing period. Pullets subjected to a constantly increasing photoperiod during the growing period come into production too early. Egg size and production suffers, and a high incidence of prolapse of the uterus may result.

After the first three days to a week, pullets raised in a light-tight windowless house may be put on a constant 8 hour light day. This lighting program will reduce problems with vices and delay sexual maturity. Pullets will come into production with more uniformly large egg size and better production when the light period is held constant or reduced to delay maturity.

Pullets raised in windowed houses are exposed to natural light conditions and must be treated differently. Those pullets hatched between September 1 and March and grown in windowed houses should receive a step-up, step-down lighting schedule. This lighting schedule is planned by determining the amount of natural daylight the pullets will receive when they reach 22 weeks of age. To this length of day is added 7 hours to give the starting length of light period. This light period is the total of

the natural light supplemented by artificial light. This light period is provided for the first week and stepped-down 20 minutes each week until the
birds reach 22 weeks of age. At 22 weeks the light is stepped-up one hour
to stimulate egg production. Each week the light period is increased by
one hour until a total of 15 or 16 hours are provided. There is no need of
providing more than a 16 hour light day.

costs of growing pullets

Many factors affect the costs of production. The most flexible cost will
be the cost of feed. Where home-grown grains are available or a range
rearing program is used, substantial savings may be made and the costs will
vary from those shown in *Table 5*.

TABLE 5

ESTIMATED COST OF
RAISING PULLET REPLACEMENTS TO 22 WEEKS*

Item	Brown-Egg Pullet	White-Egg Pullet
Chick Cost	$ .34 — .42	$ .36 — .50
Feed Cost at $.08 per pound	1.68	1.36
Brooding Cost	.03	.03
Litter and Miscellaneous	.03	.03
	$2.08 — 2.16	$1.78 — 1.92

Assumptions: Mortality 8%, 21 pound Feed Consumption for Brown-
Egg Birds, 17 pound feed consumption for White-Egg
Birds full-fed. Costs will vary depending upon quantities purchased, feed program, feed prices and other
factors.

*Interest on average investment, labor and overhead (depreciation, insur-
ance, and taxes) are costs of producing commercial pullets not normally
considered in small flock situations. The approximate costs of these items
are: interest — 7.5¢ — 10.0¢, labor — 14¢, overhead — 23¢ — 31¢.*

6

MANAGING THE LAYING FLOCK

Preparing The Laying House

The laying house should be thoroughly cleaned between flocks — this means the removal of manure, and brushing down the cobwebs and dust from walls and ceilings. Any repairs that are needed should be done at this time. Wet cleaning is best, but dry cleaning is satisfactory.

After the house is cleaned it should be sprayed with a disinfectant. The equipment should also be cleaned and disinfected at this time. Some of the common disinfectants include creosol solutions, hypochlorites and the phenol or carbolic acid preparations. Use disinfectants according to the manufacturer's recommendations.

When the house has been disinfected, the roosts, dropping boards or pits, and the nests should be painted with carbolinium or a red mite paint. It is advisable to treat the floors with carbolinium once each year. Carbolinium is not only a disinfectant but a wood preservative, and it makes future cleaning much easier. A similar cleaning and disinfecting program should be used for the brooder house and equipment.

A word of caution. Carbolinium is an excellent material but must be used correctly. Allow a minimum of two weeks for the house to dry and air out before the birds are put in. The material can be injurious to the birds' feet and eyes. After the house has had an opportunity to dry, the litter can then be added.

litter

Provide approximately 6 to 8 inches of clean, dry litter material. Although the availability and cost are factors in determining litter to be used, some of the more commonly used materials are shavings, sawdust, a combination of shavings and sawdust, ground corn cobs, and sugarcane. These are all excellent litter materials.

To help maintain the litter in a satisfactory condition, it is recommended that a built-up litter be used. Built-up litter should be started early in the fall before cold weather sets in. Start with 4"-5" of clean litter and gradually add to it. Fresh litter is added until a depth of 9"-12" is reached. It is usually not necessary to change the litter during the laying year. Removal of the wet spots from time to time is usually all that is required.

Built-up litter insulates the floor and provides warmth for the birds. Due to the decomposition and fermentation processes heat is actually produced in the litter. It also absorbs moisture from the feces.

The same litter should not be used more than one year. Infestations of both internal and external parasites can result. Then, too, it is necessary to remove the litter to properly clean the house.

When warm weather arrives, it is the practice of some producers to replace the built-up litter with new litter. This provides cooler, more comfortable, conditions during hot weather.

In cold weather it is usually necessary to stir the litter regularly to keep it from packing and to permit it to aerate and dry.

feeding the layers

Laying birds require a diet containing from 16 to 18% protein. Actually the requirements vary during the laying cycle. Protein intake needs to be higher during the early laying period, because that is when egg production peaks and the birds are still growing. As egg production diminishes, the protein requirement decreases.

Protein is expensive, so commercial producers frequently phase-feed their layers, that is, use three different protein levels during the laying period. They may start with an 18% diet and, at approximately 4 months of production, reduce the level to 16%. When the layers fall below 60-65% production the protein content of the feed is dropped to 15%.

Many factors affect feed intake of the birds. With changes in feed consumption, protein intake changes, so feeding layers is not a simple task if you are aiming for optimum production. Some of the factors which affect feed intake include:

(1) Management factors such as feed hopper and floor space.
(2) Environmental temperatures, both warm and cold.
(3) Calorie content of the diet.
(4) Variations in egg production.
(5) Flock health or stress factors.

Most small flock owners prefer to use a simple feeding program, one which utilizes one type of feed and which can be full-fed to the layers. In these situations a 16 to 17% all-mash diet is normally used.

all-mash feeding

The simplest and most foolproof approach to feeding layers is to feed a complete all-mash laying ration and keep it in front of the birds at all times. It is less bother, is adaptable to mechanical feeding and provides a more nearly balanced diet for the layers, assuming there is enough feeder space available.

grain and mash feeding

Some producers still prefer to use a grain and mash system of feeding. Grains are frequently fed in the form of so-called "scratch feed" containing corn, wheat and oats. It is often thought to be somewhat easier to maintain a high feed consumption with the grain and mash systems. Grains are also well liked by the birds and they eagerly consume them when they are provided. Some producers feed a large part of the daily allowance of grain before the birds go to roost. During cold weather this gives them a reserve supply of energy for warmth and will help keep them more comfortable while on the roost during the night. Grains are high in energy and are digested more slowly than mash, thus providing more warmth over a longer period of time. Grit must be provided when hard grains are fed.

The portions of mash to grain vary with the protein content of the mash and also the type of layer. Usually a 20-21% mash is kept before the birds at all times. Light breeds such as Leghorns, on a grain and mash program, are fed equal amounts of grain and mash, or perhaps slightly less grain than mash. Heavy breeds have a tendency to become too fat if they receive too much energy, and therefore, may receive only 40% grain and 60% mash. In some instances, if a high energy type mash is used, it would be necessary to decrease the amount of grain to 30% of the total diet. It is important to prevent laying birds from getting too fat. Excessive fat is not conducive to high egg production.

One of the problems encountered with the mash and grain system is to feed in the proportions that will provide the correct protein intake of approximately 16 percent. Production problems can frequently be traced to a low total intake of protein when a mash and scratch system is being used.

The protein content of most scratch grains is approximately 9 to 10%. If, for example, equal proportions of an 18% laying mash and scratch grains are fed, the total protein intake will be approximately 13.5 to 14%. This will not support optimum egg production — at least during the early part of the laying cycle. Most high energy complete laying mashes can be safely supplemented with grain at the rate of 1 or 2 pounds per hundred birds per day. Follow the manufacturer's recommendations.

Grain may be fed on top of the mash or in the litter if litter conditions are reasonably dry and clean. When birds scratch for grains it helps to aerate the litter and maintain it in better condition.

free choice or cafeteria style

This method uses a high-protein mash containing from 26-40% protein, with grain and calcium supplement kept before the birds at all times. The chief advantage of this method is that it enables a producer to use local grain or grain produced on his farm. It is also a simpler system than hand-feeding grain. The mash concentrate is fed in one hopper, and the grains are fed in individual hoppers. The bird actually balances its own nutrient intake, hopefully.

Other Feeding Programs

For small flock situations fresh kitchen scraps, garden products and even surplus milk can be fed to layers. Feeding of these various materials can substantially reduce the amount of purchased grain required. The amounts of these materials should be limited to what the birds can clean up in 5 or 10 minutes. If too much of this sort of material is fed to the birds, the nutritive intake can be diluted to the extent that they don't get the right amount of protein and other nutrients for body maintenance and the production of eggs. Milk should be given to the birds in plastic or enamel containers and not in galvanized containers.

Care should be taken in the choice of table scraps given the birds. Spoiled meat and materials like fruit peelings, onions and other strong flavored foods may give a bad taste to the eggs and should not be fed to the birds. Potato peelings can be fed to chickens if they are cooked. Vegetable peelings and green tops of vegetables are good. Green feeds are satisfactory for layers, and those that may be used for laying birds are similar to those recommended for growing birds. When a complete laying mash is supplemented with table scraps and other materials, grit and calcium supplements should be provided the birds. Egg shells are a good source of calcium and may be crushed and fed back to the laying flock. A simple or "Natural Laying Hen Diet" for those who want to mix their own ration, can be found on page 220.

TABLE 6

FEED CONSUMPTION PER DAY PER 100 LAYERS AND FEED PER DOZEN EGGS VARIOUS LEVELS OF PRODUCTION

Percent Production	Feed for Leghorn-Type Bird — 4 Pounds Pounds		Feed for Brown-Egg Bird — 5 Pounds Pounds		Feed for Brown-Egg Bird — 6 Pounds Pounds	
	Per Day	Per Doz.	Per Day	Per Doz.	Per Day	Per Doz.
0	15.8		18.6		21.2	
10	16.7	20.1	19.6	23.6	22.1	26.6
20	17.6	10.5	20.4	12.2	22.9	13.7
30	18.4	7.4	21.3	8.5	23.9	9.6
40	19.3	5.8	22.2	6.7	24.7	7.4
50	20.2	4.8	23.1	5.5	25.6	6.1
60	21.1	4.2	24.0	4.8	26.5	5.3
70	22.0	3.8	24.9	4.3	27.5	4.7
80	22.9	3.4	25.8	3.9	28.1	4.2
90	23.8	3.2	26.7	3.6	29.2	3.9

Based on Card, L.E., and Nesheim, *Poultry Production — Tenth Edition.*

feed consumption of layers

Feed consumption varies with the body weight and rate of production *(Table 6)*. A certain amount is used for body maintenance. This amount corresponds with the amount consumed at "0" production. As a rule of thumb to estimate feed consumption at various levels of egg production, add to the amount of feed required for maintenance ("0" production) 1 pound of feed for each 10% of egg production.

Birds eat to satisfy their energy requirements, so the feed consumption will vary with environmental temperatures. They will eat more in cold weather. To insure optimum feed consumption, provide 4 linear inches of feeder space per layer.

calcium supplements and grit

Calcium is one of the most important minerals needed in the feeding of layers. Hens need calcium for egg-shell formation. About 10% of the total weight of the egg is shell. The shell is almost 100% calcium carbonate. Hens in heavy production require calcium in relatively large quantities.

Most of the high energy complete laying mashes today contain 3 to 3½ % calcium depending upon the time of year. Under most circumstances, this appears adequate to produce eggs with sound shells.

When laying birds get well into their laying cycle and shell quality begins to deteriorate, or during periods of extremely hot weather when shell quality may suffer, additional calcium seems to be indicated. In these cases, producers frequently add extra calcium to the diet in the form of oyster shells. Some types of birds, particularly the light laying strains of birds that have relatively small calcium reserves, need calcium supplementation earlier in the laying cycle than some of the heavier types of birds. When to, and not to, feed supplemental calcium is still somewhat of a controversial subject. There are those who feel that the most satisfactory way of meeting the variable individual requirements of layers is to supplement the calcium that is in the laying mash. This is done by feeding separately hen-size oyster shell or calcite crystals, both of which are soluble.

For those feeding systems other than the complete laying mash system, a supplemental feeding of calcium is to be recommended.

In addition to a source of calcium, a hard insoluble grit should be fed in many instances. Insoluble grit, such as granite grit, maintains its sharp edges in the digestive system and in essence takes the place of teeth. It is frequently thought that it is not necessary to feed grit to layers that are fed an all-mash type of diet. It is assumed that everything the birds consume in the laying diet is already finely ground and, therefore, needs little or no further grinding. Where birds are kept in cages, this is probably true.

However, birds that are housed in a floor management system, frequently eat feathers, litter and nesting material. Under these circumstances grit, if available, enables the gizzard to grind the fibrous material so that it can be moved on through the digestive system. Any feeding program other than the all-mash system of feeding, should incorporate grit so that coarse grains can be utilized by the digestive system.

There are a number of types of grit and shell hoppers available on the market. These are usually divided into two compartments, so that both grit and the source of calcium can be fed from the same hopper. It is a relatively simple project to construct a hopper which will serve equally as well. In those situations where grit and calcium material is needed, the hopper should be kept full at all times.

watering

Birds need plenty of clean, cool water at all times if they are to produce well. Water makes up a large portion of the hen's body and is a major constituent of the egg. Water helps to soften the feed and aids in its digestion, absorption and assimilation. If hens are deprived of water for only a short time, egg production will suffer. Dirty water and watering equipment may discourage water consumption and is a potential source of disease infection. The water supply should never be permitted to freeze for an extended period. If the house temperature cannot be maintained at a level to prevent freezing, then an electrical immersion heating unit can be installed in the drinking fountain. Heating units should be thermostatically controlled. Laying birds normally drink in pounds approximately twice as much as the feed they consume. During hot weather they will consume substantially more than this amount. One inch of the trough-type waterer per bird should be provided, or one round waterer per 100 birds. One pan-type waterer is usually adequate for the small family flock. Waterers should be cleaned and filled with fresh water daily. The water consumption of layers is shown in *Table 7*.

TABLE 7

WATER CONSUMPTION OF LAYERS
BASED ON ENVIRONMENTAL TEMPERATURE

Temperature	Gallons Per 100 Layers Per Day
20-40° F	4.2-5.0
41-60° F	5.0-5.8
61-80° F	5.8-7.0
81-100° F	7.0-11.6

Source: *New England Poultry Management and Business Analysis Manual*
  — Bulletin 566

floor space

In well insulated, mechanically ventilated houses Leghorn type layers can be housed at the rate of 1¼ to 1½ square feet per bird. Heavier brown egg birds should be given 1½ to 2 square feet per bird. In those houses where gravity ventilation is used, that is, ventilation through the windows or slots, slightly more floor space should be given. The normal recommendation is a minimum of 1½ to 2 square feet for Leghorns and 2 to 2½ square feet for brown egg birds. Actually, in a well insulated house cold weather ventilation is somewhat simplified if there are more birds, since they supply more heat. When feeders and waterers are located over a dropping pit, the floor area per bird can be reduced below the amounts previously mentioned. However, with each additional reduction in floor space per bird, additional feeding, watering, nesting and roosting space must be provided. The birds should not have to travel more than 15 feet to reach either feed or water.

nests

Provide at least one individual nest or one square foot of community nest space for each four hens. Nesting material should be replenished as needed and be kept clean and dry. This is very important in preventing dirty, cracked and broken eggs. The nest should be a minimum of 2 feet from the floor or litter. If floor eggs are a problem at the onset of production, it is advisable to place the nests on the floor and gradually raise them as the birds become accustomed to using them. One of the common problems with small flocks is egg-eating. When the nests are poorly constructed or without litter, or the eggs are not gathered frequently enough, breakage occurs, and this habit is much more likely to get started.

egg gathering

Eggs should be gathered two or three times a day. Frequent gathering reduces the number of dirty eggs and improves the egg quality. It also reduces the possibility of birds developing the habit of egg-eating. Very often, it is difficult for small flock owners to gather the eggs as frequently as is desirable. Having plenty of nests, keeping them clean and with plenty of litter will help prevent broken or dirty eggs. When the flock owner has to be away during the day, and the eggs will not be collected, it is possible to have the lights come on earlier in the morning, so that part of the eggs will be laid earlier and can be gathered before leaving for the day. Once egg-eating commences it is a difficult habit to stop. Probably one of the best methods of preventing it is to debeak the birds and to increase the frequency of egg collection. Eggs not gathered during the day are also subject to freezing in cold weather.

lighting

The use of artificial light in the poultry house is not to give the hens more time to eat. It is the stimulation of the light itself that makes them lay more eggs. The light stimulates the pituitary gland through the eye. This gland, in turn, secretes hormones that stimulate the ovary of the hen to lay eggs.

Laying birds need about 14 hours of light daily. In the northern part of the United States, darkness exceeds the amount of daylight during many of the fall and winter months. Beginning on August 15th, the birds need to receive extra light to give them a 14-15 hour day. During the months of November, December and January, when the days are shortest, they will need 5 hours of supplemental light. The lights may be turned on in the morning or in the evening. A combination of both morning and evening lights may be used, as long as the birds have at least 14 hours of light each day. With small flocks it is a task to turn the lights off each night, or to get up early enough to turn them on in the morning. The best way to solve this problem is to use a time clock which automatically turns the lights off and on at the desired time. This generally gives satisfactory results. As soon as the natural day length reaches 14 hours again, the lights can be discontinued.

The lights should be ordinary 40-60 watt incandescent bulbs. To get the most efficient use of the light wattage a reflector can be used. This will frequently permit the use of lower wattage bulbs. One light fixture should be installed for each 200 square feet of floor area or less. A distance of 10 feet between lights generally gives good distribution of light for the birds. A rough rule of thumb for light intensity is one-foot candle at the feeder level. One bulb watt per 4 square feet of floor space will usually provide 1 foot-candle if the bulb is 7 to 8 feet from the litter and has a reflector. During the first week or two of the lighting period, it may be necessary to catch and place some of the birds on the perches if they do not get on the roost of their own accord when it's time for *lights out*.

ventilation

During cold weather all of the ventilation should be from one side of the poultry house. Preferably the south. Windows that drop down from the top or tip in are best for winter ventilation, in the absence of fan ventilation. The windows should be adjusted according to the weather. When the house tends to be stuffy, and the ammonia fumes are strong, the house needs more ventilation. The house should never be closed tight, even on cold nights. It is always well to leave at least some of the windows slightly cracked.

In the warm, summer months or in warm climates, ventilation will help cool the house. If you can open up both sides of the house, it will increase the air movement, help keep the birds cool and maintain better egg production.

rodent and bird control

Rats and mice like chicken feed. If there is a place for them to hide in the chicken house they will do it. They enjoy living in piles of refuse, the double walls of buildings, old stone foundations, in dropping pits, under deep litter or under the floors, and many other hiding places. Good housekeeping will help to reduce the problem with rats and mice.

Anticoagulant rat poisons and other types of poisons have been very helpful in getting rid of rat problems. For anticoagulant-type poisons to be effective, rodents must consume it for several days. If birds are housed in an old building that has plenty of hiding places, it may be best to set up a permanent bait station or stations, and keep bait in them at all times.

It is desirable to screen wild birds out of the chicken house, if possible. Wild birds are vectors of some of the common diseases and parasites of poultry.

culling and selection

Good production-bred chickens will lay well for approximately a year if they are well managed. The culling of healthy birds that go out of production for just a short time is not advisable, unless a chicken dinner is in order. After the birds have laid for nearly a year, the non-layers may be culled out, dressed and used at home or sold. Usually the entire flock should be replaced after 12 to 14 months of production. On occasion it is necessary — and advisable — to sell hens that are still laying at a reasonable rate, to make room for the new flock of pullets.

In culling, there are two distinct things that one has to learn. One is how to distinguish between a layer and non-layer. The other is to determine with a reasonable degree of accuracy, how long the bird has been either in or out of production. After one has acquired the ability to do these two things, he can do a reasonably good job of culling a flock of layers.

With the beginning and termination of laying, changes take place in the head, the abdomen and the vent. To determine whether or not a layer is in production, all three of the body parts should be carefully examined to make an accurate judgment. The examination of one of these parts alone will not provide the whole answer.

When birds are in laying condition their combs and wattles are enlarged, bright red and waxy in appearance. The vent is large and moist and the pubic or pin bones are wide apart to physically permit the egg to be laid. The spread of the pubic bones of the laying bird will be about three fingers wide. The abdomen will be soft and pliable.

When birds are not laying, the combs and wattles may be small, pale and scaly in appearance, the vent dry and puckered. The pubic bones may be close together (one or two fingers width) and the abdomen hard and shallow.

The characteristics of layers and non-layers and high and low producers are presented in *Table 8*.

To determine how long a bird has been in production, or out of production, one must consider two factors, namely, the degree of bleaching of pigment and the molt.

TABLE 8

CHARACTERISTICS OF LAYERS AND NONLAYERS

Character	Laying Hen	Nonlaying Hen
Comb............	Large, red, waxy, full	Small, pale, scaly, shrunken
Wattles..........	Large, prominent	Small, contracted
Vent	Large, moist	Dry, puckered
Abdomen......	Full, soft, velvety, pliable	Shallow or full of hard fat
Pubic bones.	Flexible, wide open	Stiff, close together

CHARACTERISTICS OF HIGH AND LOW PRODUCERS

Character	High Producers	Low Producers
Vent	Bleached, large, oval, moist	Yellow, dry, round, puckered
Eye Ring........	Bleached	Yellow-tinted
Beak..............	Bleached or bleaching	Yellow or growing yellow
Shanks..........	Pale yellow to white, thin, flat	Yellowish, round, full
Head	Clean-cut, bright red, balanced	Coarse or overrefined, dull, long, flat
Eyes..............	Prominent, bright, sparkling	Sunken, listless
Face..............	Clean-cut, lean, free from yellow color and feathers	Sunken or beefy, full, yellowish, feathered
Body	Deep	Shallow
Back..............	Wide. Width carried out to pubic bones	Narrow, tapering, pinched
Plumage........	Worn, dry, soiled	Smooth, glossy, unsoiled
Molt..............	Late molter	Early molter
Carriage	Active and alert	Lazy and listless

Source: *Culling For High Egg Production, Vt. Agr. Ext. Serv. Circular 115Ru.*

Pigmentation

The yellow pigmentation found in all yellow-skin breeds and varieties of chickens is a pigment called xanthophyl. Yellow corn, a major constituent of the poultry diet, is the principal source of this pigment. If a pullet has been properly fed when it comes into the laying house, she will carry a considerable amount of pigmentation in all parts of her body. When she commences to lay, the pigment present in the skin and in other parts of the body is gradually lost. The reason for this loss is that the pigment is diverted to the yolks, giving them their yellow color. As long as the bird is in production, it continues to lose this yellow color from the various parts of its body. Hens that are thoroughly bleached out are usually high producers. When the bird stops production, the pigment reappears in the various parts of the body. The return of pigmentation is one of the first clues that a bird is out of production.

There is a very definite order of bleaching of the body parts as the birds commence to lay — the vent, the eye ring, the ear lobe, the beak, the shanks and feet. The order of bleaching and the approximate time required to bleach the various parts of the body are given in *Table 9*.

TABLE 9

ORDER OF BLEACHING AND TIME REQUIRED WHEN BIRDS ARE IN CONTINUOUS PRODUCTION

Body Part	Time Required
Vent	Few Days
Eye Ring	2 to 3 Weeks
Ear Lobe	3 to 4 Weeks
Beak	6 Weeks
Shanks	2 to 5 Months

vent

The first noticeable change occurs in the skin around the edges of the vent. Within a few days after production commences, the yellow color around the vent disappears.

A yellow vent indicates that a bird is not laying. A whitish, pinkish or bluish white vent indicates that a hen is laying.

the eye ring

The eye ring starts to bleach soon after the vent. In 2 to 3 weeks after the onset of production, the eye ring usually loses its yellow pigment. By that time, the bird has probably laid 10 to 12 eggs.

the beak

The yellow color leaves the beak next. It leaves the base of the beak first, and the fading continues toward the tip. It takes approximately 6 weeks of continuous production for complete bleaching. By that time usually 30 to 40 eggs have been laid.

The lower beak bleaches more rapidly than the upper beak. Fading of color can be readily seen in the lower beak. When the dark pigment "horn" appears on the upper beak, the lower beak can be used as a basis for judging the degree of pigmentation loss. Barred Plymouth Rocks and Rhode Island Reds frequently carry this dark pigment.

the shanks

The bleaching of the shanks occurs last and is an indicator of long-term production. Bleaching of the shanks takes from 4-5 months of continuous production. She must lay from 120-140 eggs to be completely bleached. It is not possible to pick a bird as a layer or non-layer on the basis of bleached or yellow legs unless she has been in production for 4 or 5 months.

When the bird stops laying, pigmentation returns to the different parts of the body in the same order that it bleached out.

By observing the degree of pigmentation, one can tell rather accurately how well a bird is laying. It should be taken into consideration that the feeding program, breed, and flock health are factors which can also affect pigmentation.

The Molt

Once each year birds renew their plumage. This process of replacing old feathers with new ones is called molt. Hens usually go through their annual molt in the late summer, the fall, or early winter months. Factors which determine the time of the molt are (1) time of the year that the bird was hatched, (2) the individual bird or breeding and (3) management stresses to which the bird is exposed. Where bleaching is most helpful in determining layers from non-layers during the first eight or nine months of production, the molt is most useful during the last several months of production.

When a bird starts her molt she goes out of production and, normally, will not come back into production until shortly before the molt is completed or just after it is completed. The pattern depends, to a certain extent, upon the type of management and feeding program to which she is exposed. Some laying birds start their molt early in the fall after eight or nine months of production and are called early molters. Others lay for 12 or 15 months before they molt and stop production. In fact, some of our modern day strains of layers have to be force-molted to get them out of production. This procedure provides them with an opportunity to rest, renew their feathers and return to production. Hens that complete at least 12 months of production before molting are referred to as late molters and are the most desirable birds to have in the flock.

When using the molt to cull or select hens, the primary (flight) feathers of the wing are used. There are usually ten of these feathers, arranged in a group, extending from the short axial feather in the center of the wing to the tip.

The fact that the primary feathers are dropped and renewed in a definite order makes the use of the molt a fairly accurate means of estimating how long a bird has been out of production and how long it will take her to go through her complete molt. It is also possible to determine whether she is a slow molter or a rapid molter by observing the way in which she has dropped and is renewing the primary feathers. A typical slow-molting bird will drop one primary at a time over an extended period, whereas a rapid molting hen will drop more than one at a time and drop them more frequently. In some of the better strains of laying birds, the hens will drop a group of several primaries and then, very shortly after, drop more and thus go through the complete molt in a relatively short period of time and be back in production. Approximately 6 weeks are required to grow a primary feather. Usually the slow molter stops production before or at the time they start their wing molt. They very seldom lay through a molt. Thus, the slow molter that drops one primary at a time will be out of production for many weeks and would be a like candidate to be culled from the flock. Many of the rapid molting birds will lay for a period of time after they start to renew their primary feathers. Some of the good laying strains of birds will renew half or more of their primaries before they finally stop laying. *Figure 14* shows the different stages of molt in fast and slow molting birds.

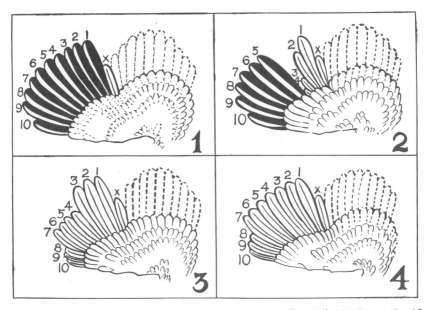

(Figure 14. Wings during different stages of molt. Top left (1) shows the 10 old primary feathers (black), and the secondary feathers (broken outline), separated by the axial feather (x). Top right (2) shows a slow molter at six weeks of molt, with one fully grown primary and feathers 2, 3, and 4 developing at two-week intervals. In contrast, 3 (lower left), a fast molter, has all new feathers. Feathers 1 to 3 were dropped first (now fully developed); feathers 4 to 7 were dropped next (now four weeks old); and feathers 8 to 10 were dropped last (now two weeks old). Two weeks later (4, lower right), feathers 1 to 7 are fully grown. Fast molt took 10 weeks, compared to 24 weeks for slow molt (South Dakota Extension drawing.)

force molting

Toward the end of the laying cycle, egg production is at a relatively low level. Interior and exterior egg quality are poorer, particularly if the tail end of the cycle coincides with hot weather. At this point some producers force molt or recycle the flock if it has been a good flock.

Force molting gives the layers a rest for about 8 weeks. After the rest, egg production increases but not to the level of the pullet year. It may reach 88-90% of the pullet year production level. Egg quality, both exterior and interior, recovers substantially.

Recycled layers usually lay profitably for only 6-9 months. Egg size is large, feed consumption and mortality is higher and the overall egg quality is lower than during than during the pullet year.

The big reason for force molting is to beat the high cost of pullet depreciation which may be 8 to 10 cents per dozen depending upon the pullet cost and fowl or salvage value. If force molted birds are to be profitable to the commercial producer, there must be a market for the extra-large eggs. The profitability of the practice is still under investigation. Most producers feel that it is better to replace the flock at the end of the first laying cycle.

Force Molt Procedure*

Day 1   Decrease lights to 8 hours per day in light-tight houses or no-natural-light in window houses. Remove all feed and water.**

Day 3   Provide water

Day 8   Provide growing mash — **40%** normal consumption level.

Day 22   Restore lights to pre-molt level and full feed laying ration.

          **   *An alternative method, particularly during periods of hot weather, is as follows:*

Day 1   Decrease lights to 8 hours per day in light-tight houses or no-artificial-light in window houses. Remove feed but provide water free choice.

Day 8   Provide growing mash — 40% normal consumption level.

Day 22   Restore lights to pre-molt level and full feed laying ration.

 * From New England Poultry Management and Business Analysis Manual, Bulletin 566.

broodiness

Poultry geneticists have done a great deal to eliminate the natural tendency for laying birds to become broody. There will, however, still be an occasional bird in floor management systems that will become broody. If the bird is permitted to stay in the nest, she will remain broody and will not eat and drink properly during this time, and a loss in body weight and egg production will result. If broody hens are removed from the nest as soon as they are discovered, the loss of egg production can be reduced substantially.

The best procedure for breaking up broody birds is to place them in a broody coop — constructed with a wire or slatted bottom to discourage setting and inactivity. The broody coops are usually suspended from the ceiling in the pen or outside the pen, and the birds are provided with plenty of feed and water. Birds exposed to this type of treatment can usually be put back with the flock in four or five days.

flock health

Flocks that are well managed and fed a well-balanced diet do not frequently experience disease problems. If one bird dies now and then it is probably nothing to become alarmed about. If a number of birds become sick, respiratory symptoms appear, or they look droopy, go off feed or stop laying — it is then time to find out why.

If a disease outbreak occurs, the cause should be determined as soon as possible. First signs are frequently a change in feed and water consumption. If you do not recognize the disease or parasite, take or send some of the live birds showing symptoms to the nearest poultry diagnostic laboratory. Poultry diagnosticians prefer live birds showing symptoms. Dead birds sent to the laboratory should be kept cool so that they do not arrive at the laboratory in a decomposed condition. It is helpful to poultry diagnosticians to know the flock history. Relate the symptoms you have observed in the flock, the number of birds affected, the number of birds that have died, source and size of the flock, feeding program, age of the flock and any other information that you think will be helpful to them.

To control disease on the farm, incinerate dead birds, or put them into a disposal pit and remove all obviously sick chickens from the flock. It is sometimes best to kill them and dispose of the carcasses. If, however, the birds are to be treated, put them in a separate pen or hospital pen — as far away from the other birds as possible.

The old adage — "an ounce of prevention is worth a pound of cure" — certainly applies to poultry flock health. If you are successful in buying or raising good healthy pullets, they are properly vaccinated, placed in cleaned and disinfected quarters, given enough feeder, water and floor space and are fed a well balanced diet, you are well over the hurdle with regard to disease or production problems.

Chickens are creatures of habit; any sudden changes can cause them stress. This is true of layers in peak production and those nearing the end of their laying cycle. Any change in feed or management should be done gradually to avoid stress problems.

Birds should be vaccinated for such diseases as infectious bronchitis, Newcastle disease, fowl cholera, laryngotracheitis and fowlpox, if these are a problem in your area. A good laying-flock manager checks for both internal and external parasites and immediately commences control measures as indicated. The screening of windows, doors and other points of entry for wild birds will help to prevent some disease and parasite problems.

The immediate disposal of dead birds, by burying or incinerating is recommended. Check local and state ordinances concerning the disposal of dead birds in your area. Some types of incinerators are not approved in certain states.

Cannibalism, feather pulling and egg eating are habits which are sometimes hard to stop once they get started. Birds commence these vices usually as a result of faulty management or environment. This trait is more common in some breeds of chickens than in others. Different types of feed may also cause these problems. Crumbles and pellets can bring on the picking problem when fed instead of mash. The birds fulfill their intake needs more quickly on pellets and have time to get into mischief. Proper debeaking is usually the answer when those problems commence. Actually the birds should be properly debeaked by 16 weeks of age so that it doesn't have to be done after the birds come into production. Debeaking after production commences can cause a production slump.

Frozen combs are one of the small flock hazards in cold climates. The combs of some breeds and varieties, such as the Single Comb White Leghorn, and the males of many breeds, have large combs. These are quite easily frozen in cold housing and can cause lowered egg production and a reduction of egg fertility in mated flocks. When severely frozen, the top of the comb becomes discolored and eventually may slough off. Naturally there are sore heads, and feed consumption may be discouraged by grills or reels on feeding and watering equipment.

This problem can be avoided by dubbing the birds. The best time to do this is at the hatchery or the farm at day of age. It is done with a pair of manicure scissors. The comb is cut off close to the head, giving a closely cropped comb at maturity, one which is less likely to be frozen or injured. White Leghorns that are to be put in laying cages are frequently dubbed to avoid injury which may occur from the cage wires. Dubbing can be done to older birds, at which time both the comb and wattles are cut with dull shears. Considerable bleeding occurs in older birds.

records

Record keeping is not only interesting but necessary if you want to know what the flock is doing and the costs and returns involved. Records need not be complicated but merely contain information on daily egg production, mortality, culling, feed consumption, the quantity and value of poultry and eggs eaten and sold.

Records may be kept on a calendar or on a pen record obtainable from most feed companies. You can make your own pen record. At the end of the month the information can be transferred from the pen record to a permanent record. These records should be analyzed each month and at the end of the year. Records should be filled in each day to be accurate. Without good records it is hard to tell if you should continue in business or buy your eggs and poultry at the store.

costs of egg production

Budgets for small flocks can vary substantially from the budget information in *Tables 10* and *11*. Overhead costs, interest and labor are frequently not considered in the small flock situation. Bird depreciation may vary considerably. The fowl may be processed and used at home, or sold, and would thus be valued at the consumer price instead of the farm price for fowl which is relatively low. Feed price is another variable. Small flock owners normally buy their feed in small quantities at a price well over that paid by the commercial producer. Feed prices also will depend upon the feeding program, the geographical area and other factors.

The two budgets are based on the following assumptions:

Birds housed at 22 weeks of age, sold at 73 weeks of age. (51 weeks of production).

Annual mortality   12%
Eggs per hen housed   221

Feed per 100 daily

Brown-egg birds 26 pounds (92.8 pounds annual)
White-egg birds 23 pounds (92.1 pounds annual)

TABLE 10

COST PER DOZEN ANALYSIS — WHITE EGGS

Item	Cost (Cents)
Feed (4.19 pounds) at 8.00 per cwt.	33.52
Bird Depreciation	10.00
Overhead	1.77
Other	1.68
Interest	0.71
Labor	1.74
Total	49.42

Based on information from the *New England Poultry Management and Business Analysis Manual, Bulletin 566.*

TABLE 11

COST PER DOZEN ANALYSIS — BROWN EGGS

Item	Cost (Cents)
Feed (4.73 pounds) at 8.00 per cwt.	37.84
Bird Depreciation	8.50
Overhead	2.26
Other	2.01
Interest	0.87
Labor	1.78
Total	53.26

Based on information from the *New England Poultry Management and Business Analysis Manual, Bulletin 566.*

7

THE PRODUCTION
OF FERTILE EGGS

A rooster is not needed in the laying flock. The hens will lay just as well without him. Some individuals do, however, like to mate their flock of layers to merely produce fertile eggs. Others plan to perpetuate a breed or strain of bird as a hobby, for exhibition purposes, or for a number of other reasons.

If flocks are mated for the production of eggs for hatching, it is well to check on state regulations. Some states require that all hatching egg flocks be blood-tested for Salmonella Pullorum and Salmonella Typhoid. Blood testing is supervised by employees of the official state agency that administers the National Poultry Improvement Plan in the state. For further information on blood test requirements, check with your State Department of Agriculture, State Extension Poultryman or County Extension Agent.

As is the case with other laying flocks, careful feeding and management, good housing and proper equipment, all apply to hatching egg production. Good feeding and management are important if one is to get a good hatch of healthy chicks from the eggs.

Preliminaries

selecting stock

Select a source of stock that is healthy and bred to meet the specifications you have set, whether it is feather color, body size, egg production, meat production or a combination of these characteristics.

mating

A hatching egg flock of Leghorn or Leghorn-type birds should contain about 6-7 males per 100 pullets. Heavier breeds, such as the Rhode Island Red and the Plymouth Rocks are mated at the rate of 8 males per 100 pullets. The ratio of males to pullets in some of the commercial hatching egg flocks may vary from this ratio substantially, due to anticipated mortality, morbidity or other special problems with the males.

If you are trying to perpetuate a breed, the breeding stock should be selected to conform to those characteristics, desirable for that breed or variety, as outlined in the American Standard of Perfection. The American Standard of Perfection is issued by the American Poultry Association. It describes the various classes, breeds and varieties of poultry which are recognized as Standard-Bred. Primary breeders select for numerous economic characteristics, including production, production efficiency, egg size, egg quality, liveability and many others.

If yearling hens are to be mated, force them out of production to permit approximately an eight week rest before they commence laying hatching eggs. The same force molting procedure is used as outlined in *Chapter 5*.

The flock should be mated at least 2 weeks before hatching eggs are to be saved.

feeding and management of breeders

In addition to producing eggs efficiently, breeding stock should produce eggs that hatch well and produce vigorous chicks. Breeder flocks should be fed a diet that has additional vitamins and minerals for high fertility and hatchability. For best results then, the breeding flock should be fed a specialized diet, formulated for hatching egg production. These are known as breeder rations.

The management of a breeding flock is very similar to a market egg flock. Most of the management recommendations given in the preceding chapter on Laying Flock Management, therefore, need not be repeated in this chapter. Only the basic differences between breeder flock and laying flock management will be mentioned.

One of these basic differences is the management system required. It was mentioned in the preceding chapter that there are several management systems available for laying birds. Among those mentioned were the cage system, the combination slats and litter, the all slat or the combination wire and litter, or all wire, or the conventional litter management approach.

Cage management systems for breeding birds are relatively uncommon. This is because of the difficulty in designing cages so that birds can mate with good results. Where cage systems are used it is usually necessary to artificially inseminate the females. This is not a difficult task where small flocks are involved, but is much more time consuming than the conventional floor pen matings.

While some of the other management systems have been used with some degree of success, there are a number of problems involved with systems other than the littered floors. This is particularly true with the heavier birds that appear to be inherently lazy and don't want to lay well in the nest. For example, the all slat floors or even the slat and litter management systems have often produced a high percentage of floor eggs or eggs layed on the slats. This can cause dirty eggs, broken eggs, and the loss of eggs through the slats. It is probably best not to use wire floors for mated flocks, particularly the heavy breed flocks, because of the incidence of foot and leg problems and the relative difficulty of the birds to mate.

lighting the breeders

Light is important not only for the stimulation of egg production in breeding flocks, but it also increases the semen output of the males. Lighting schedules were outlined in *Chapter 6, Managing the Laying Flock*. Light schedules vary with the type of house and very careful attention to the lighting schedule should be given.

Hatchability of fertile eggs is affected by many factors. Shell quality is one of these factors. The longer the bird is in production, the poorer the shell quality. Shell quality is affected not only by the length of time the birds have been laying, but also the time of year, temperature, diet and genetics. Birds that have been force molted and are in a second year of egg production invariably show shell quality deterioration much more rapidly than during the first laying cycle.

Tremulous or loose air cells also affect the hatchability of eggs. Use care in handling eggs to prevent tremulous or loose air cells. Eggs that have been cracked very seldom hatch, so care should be taken to produce as few cracks as possible and to make sure that cracked eggs are not set in the incubator.

Eggs
gathering and care of hatching eggs

Hatching eggs should be gathered 3 to 4 times a day. When temperatures are extremely hot or cold, they should be gathered more frequently. Eggs should not be left in the nest overnight.

Eggs should be cleaned as soon after gathering as possible. Unless they can be washed properly, they should not be washed but rather dry-cleaned by sanding or buffing. Hand-buffing devices are available for this purpose.

Eggs should be washed at a temperature of 110°-115° F. in water containing an approved detergent-sanitizer. They should not be immersed in the wash water for more than 3 minutes. The eggs should be dried and moved to cold storage. Improperly cleaned eggs frequently become contaminated and may explode during incubation.

The eggs should be stored at a temperature of 50-60° F. Lower storage temperatures may reduce hatchability. The temperature of the average household refrigerator runs below that recommended for storing hatching eggs. For best results, the relative humidity should be 75%.

Eggs that are cracked, have thin shells, shells with ridges or that are excessively dirty or abnormal in size or shape should not be kept for hatching. Excessively large or small eggs are often infertile or won't hatch and should not be set in the incubator.

For best results, hatching eggs should not be stored for more than 10 days to 2 weeks before they are set. They should be stored in filler flats or egg cartons with the small end down. Eggs that are to be held for several days prior to hatching should be held in a slanted position (approximately 35°) and turned at least twice each day. One simple method of turning eggs is to prop one end of the egg carton or case up on an object at an angle of about 35°. Then just shift ends with the container at least twice each day.

Incubation: types of machines and operating temperatures

There are two types of incubators in use today. These are the forced-draft and the still-air machines. The one most frequently used in commercial hatcheries is the forced-draft machine. This type of incubator has fans which force the air through the machine and around the eggs. For most types of eggs the incubator temperature of the forced-draft machine is set at 99½° F. to 99¾° F.

Most of the still-air incubators in use today are quite small. They are made to hold from one to 100 or more eggs. Still-air machines do not have fans. They depend upon gravity ventilation through vents on the top and bottom of the machine. The operating temperature of the still-air machine is higher than for the forced-draft incubator. It ranges from 101½ to 102¾° F. depending on the type of eggs being incubated. Actually good results may be obtained using a constant 102° F. temperature. The temperature may vary between 100° and 103° F. without hurting the embryos if the temperature does not stay at these extremes.

humidity

Humidity in the incubator varies from 83° to 88° F. (wet bulb thermometer) depending on the type of eggs. This level of humidity is maintained until the last 3 or 4 days before hatching, and then is increased to 90-95° F. (wet bulb).

The humidity in small incubators is provided by moisture evaporation pans. The evaporation pan has to be kept full at all times. When adding water, add warm water (110° F.) so as not to reduce the incubator temperature for an extended period of time. If the humidity in the machine needs to be elevated, a large sponge, placed in the evaporation pan, will help to increase the moisture level. More pan space also helps.

During the incubation period (19 days), eggs should lose about 11% of their original weight through evaporation. A loss of more than this amount is detrimental. The amount of evaporation is controlled by proper humidity. The relative humidity level for chicken eggs should be 50-55% for the first 18 days (83°-87° F. wet bulb reading) and 65% (89°-90° F. wet bulb reading) during the last three days. If humidity is too low, close the vents part way, or add to the moisture evaporation surface. If humidity is too high, permit more ventilation.

Figure 15 shows the approximate size of the air cell when evaporation is normal. By candling the egg at various stages of incubation, this can be used as a rough guide to control humidity in the absence of a wet bulb thermometer. The candling procedure and candling light are covered in the section on candling.

7TH DAY
14TH DAY
18TH DAY

(Figure 15. Air cell size at various stages of incubation.)

ventilation

Ventilation is very important in the incubator. Without it the embryos will suffocate. For machines with adjustable vent opening, the openings are usually cracked open at the start and are gradually opened to permit more ventilation toward the end of the incubation period. Follow the manufacturer's directions, or if using a homemade machine, experiment until the best results are obtained. The room where the incubator is located should be ventilated, yet not drafty.

turning the eggs

While eggs are incubating they should be turned at least four times a day. This is to prevent the embryo from sticking to the shell membrane. Large machines may include automatic or mechanical turning devices. If the eggs are not turned during the night, they should be turned late in the evening and early in the morning. If the eggs are set on their ends they should be slanted at about 30°. If they lie flat in the tray they should be turned from one side to the opposite — in other words, 180°. One method of making sure that all eggs are turned is to put an "X" on one side of the egg and an "O" on the other, in pencil. When turning the eggs make sure all the "X's" or "O's" are on top. Discontinue turning after the 18th day for chicken eggs or 3 days prior to hatching for eggs of all other species.

candling

Normally, the eggs are candled three days prior to hatching and the infertiles removed. Candling is done in a dark room using a special light. *Figure 16* shows a candling light which can be easily made at home. If the eggs are infertile, they will appear clear before the candling light. The fertile eggs will permit only light through the large or air cell end of the egg. The rest of the egg will be black or very dark in color. The eggs may be candled earlier in the incubation period. At 72 hours the early fertiles will have the typical blood vessel formation, looking much like a spider.

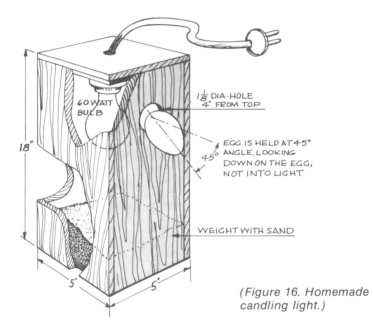

(Figure 16. Homemade candling light.)

After the eggs hatch, leave the young birds in the incubator for approximately 12 hours until they are dried and fluffy before removing them. The young can survive without food and water for 72 hours, but the sooner they are put on feed and water, the better.

Table 12 contains incubation trouble shooting information.

Table 13 lists the incubation period for eggs of various species.

TABLE 12

INCUBATION TROUBLE CHART

Symptom of Trouble	Probable Cause	Suggested Remedies
Eggs Clear No blood ring or embryo growth.	1. Improper mating.	1. 8 to 10 vigorous males per 100 birds.
	2. Eggs too old.	2. Eggs set within 10 days after date laid.
	3. Brooding hens too thin.	3. Keep hens in good flesh.
	4. Birds too closely confined.	
Eggs Candling Clear But showing blood or very small embryo on breaking.	1. Incubator temp. too high.	1. Watch incubator temp.
	2. Badly chilled eggs.	2. Protect eggs against freezing temp.
	3. Breeding flock out of condition (frozen combs, chicken pen, etc.)	3. Do not set eggs from birds with frozen combs or with contagious diseases.
	4. Low vitamin ration.	4. Feed fish oil and alfalfa.
Dead Germs Embryos dying at from 12-18 days.	1. Wrong turning.	1. Close temp. regulation.
	2. Lack of ventilation.	2. Plenty of fresh air in incubator room and good ventilation of machines.
	3. Faulty rations.	3. Feed yellow corn, milk, alfalfa meal and fish oil.
Chick Fully Formed But dead without pipping.	1. Improper turning.	1. Turn eggs 4 times daily.
	2. Heredity.	2. Select for high hatchability.
	3. Wrong temperature.	3. Watch incubator temp.
Eggs Pipped Chick dead in shell	1. Low avg. humidity	1. Keep wet bulb temp. from 85-90° F.
	2. Low avg. temperature.	2. Maintain proper temp. throughout hatch.
	3. Excessive high temp. for short period.	3. Guard against temp. surge.

81

Sticky Chicks Shell sticking to chick.	1. Eggs dried down too much	1. Carry wet bulb temp. at 85° F. between hatches.
	2. Low humidity at hatching time.	2. Increase wet bulb reading to 88-90° when eggs start pipping.
Sticky Chicks Chicks smeared with egg contents.	1. Low avg. temperature.	1. Proper operating temp.
	2. Small air cell due to high avg. humidity.	2. Increase ventilation and lower humidity.
Rough Navels	1. High temp. or wide temp. variations.	1. Careful operation.
	2. Low humidity.	2. Proper humidity.
Chicks Too Small	1. Small eggs.	1. Set nothing under 23 oz. eggs.
	2. Low humidity.	2. Maintain proper humidity.
	3. High temperature.	3. Watch incubator temp.
Large Soft Bodied Chicks	1. Low avg. temperature.	1. Proper temperature.
	2. Poor ventilation.	2. Adequate ventilation.
Mushy Chicks	1. Navel infection in incubator.	1. Careful fumigation of incubator between hatches.
Short Down on Chicks	1. High temperature.	1. Proper temperature.
	2. Low humidity.	2. Careful moisture control.
Hatching Too Early With bloody navels	1. Temperature too high.	1. Proper control of temp.
Draggy Hatch Some chicks early, but hatch slow in finishing.	1. Temperature too high.	1. Proper operation.
Delayed Hatch Eggs not starting to pip until 21st day, or later.	1. Avg. temp. too low.	1. Watch temp., check thermometers.

Crippled Chicks	1. Cross beak — heridity.	1. Careful flock culling.
	2. Missing eye — abnormal.	2. Matter of chance.
	3. Crooked toes — temp.	3. Watch temperature.
	4. Wry neck — nutrition (?)	4. Not fully known.
Excessively Yellow	1. Too much formaldehyde, fumigation.	1. Follow directions on fumigation program.

Source: *Embryology and Biology of Chickens, University of Vermont*

TABLE 13

INCUBATION PERIODS FOR VARIOUS SPECIES

Species	Days
Chicken	21
Duck	28
Muscovy duck	33-37
Turkey	28
Goose	30-32
Guinea	26-28
Pheasant (Ring Neck)	23-24
Mongolian Pheasant	24-25
Ostrich	42
Pigeon	16-20
Japanese Quail	17
Bobwhite Quail	23
Peafowl	28

Constructing A Small Incubator

A small still-air or forced-draft incubator can be constructed quite easily and inexpensively *(Figure 17)*. There are several ways to make them. If the incubator is to be used for demonstration purposes, the top or one side can be made of glass, so that the hatching process can be observed. Some excellent, still-air incubators have been constructed from styrofoam coolers. Others have been made from cardboard boxes, or fabricated from plywood.

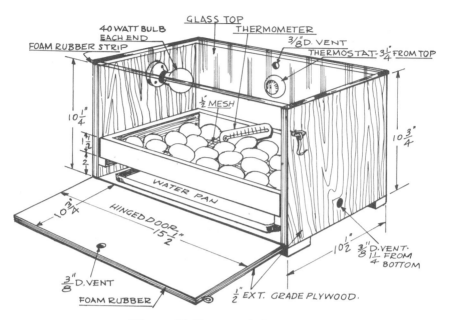

(Figure 17. Homemade incubator.)

Some commercial concerns offer incubator plans and kits for sale, and, of course, some companies also sell incubators. A list of these concerns selling incubators and incubator parts may be found in the section, *Sources of Supplies and Equipment.*

natural incubation

The first requirement for hatching eggs naturally is a good setting hen. Many of our modern-day chickens have had the broodiness characteristic bred out of them. Some of the breeds that are usually good brooders are the Rhode Island Reds, Plymouth Rocks, Wyandottes and Orpingtons. Don't use birds with long spurs. They are likely to trample the chicks. Select a calm bird. She will be less likely to break eggs in the event that she becomes frightened.

Provide a suitable nesting box, one which is roomy and deep so that she will have plenty of room to change her position, turn the eggs, and be comfortable. Select a spot for the nest where it will be quiet, away from dogs or other predators and disturbances. Darken the nest area and provide feed and water close by. The bird should be free of lice or mites before setting her.

A medium sized hen can cover 9 or 10 eggs, a large one 15 eggs. The same birds can cover only 6 or 7 turkey eggs, 9 to 11 duck eggs and 4 to 5 goose eggs. Bantams normally cover 8 or 9 of their own eggs.

It may be advisable to try the bird on dummy or artificial eggs for a few days to see if she is a persistent brooder. Take her off the eggs a few times. If she gets back on the nest each time, then put the hatching eggs under her.

Rhode Island Reds*

Single Comb White Leghorns*

Sex-Sal chicken
(Popular brown egg
New England breed)

*Source: *Poultry Tribune, Mount Norris, Illinois*

8

MEAT BIRD PRODUCTION

Birds grown for meat include, broilers, roaster, capons, turkeys and waterfowl. Turkeys and waterfowl seemed important enough in themselves to warrant separate chapters. Thus, they are covered in *Chapters 9* and *10*.

Most meat type birds, and particularly broilers, were once produced as a by-product of chick replacements for egg production. Chicks were purchased as straight run, that is approximately 50% cockerels and 50% pullets. The cockerels were separated from the pullets at about 10 to 14 weeks of age and marketed as broilers or fryers or carried on to heavier weights as roasters. Males that were to be used as capons were taken out, or separated from the pullets, at about 5 or 6 weeks of age, surgically caponized and carried through to marketing time. Today these cockerels are destroyed at the hatchery, in most instances, because they are relatively inefficient producers of meat.

For top performance, birds that are to be grown for meat should be selected from those sources that have outstanding broiler birds. Today's broiler chick is a very highly specialized bird. It is not a good egg producer but very efficient as a producer of meat. Broiler chicks are usually crossbreds or hybrids and the parent stock that produces the chicks are usually of White Cornish and White Plymouth Rock breeding. White feathers are preferred in meat birds for the ease of dressing and for better carcass appearance. The Cornish White Rock crosses give chicks that have white or predominently white feathers and also produce chicks with meaty breasts, large legs and with rapid, efficient growth characteristics. The dressed birds have excellent conformation and market appearance.

Breeders are continually improving their stocks. When buying a broiler chick or any chick that will be used for meat production, check the growth rate and feed conversion of available stocks in your area. Contact your State Poultry Extension specialist or County Agent. It pays to start with the best chicks one can get. Chick cost is a small portion of the total costs of producing a pound of poultry meat.

Table 14 presents information on broiler growth, feed consumption and feed conversion. Water consumption data is presented in *Table 15*. The costs of broiler production are shown in *Table 16*.

Most all meat type birds are sold as day-old chicks. A possible exception is the started capon which may be purchased at 4-5 weeks of age from a hatchery or dealer.

White Plymouth Rock chickens

Source: *Poultry Tribune, Mount Norris, Illinois*

TABLE 14

STANDARDS FOR BROILER GROWTH AND FEED CONSUMPTION

Week	Live Weight — Pounds			Gain over preceding week — Mixed sexes (pounds)	Feed per broiler for the week — Mixed sexes (pounds)	Feed to date — Mixed sexes (pounds)	Feed conversion to date — Mixed sexes*
	Cockerels	Pullets	Mixed Sexes				
1	.27	.25	.26	.17	.29	.29	1.11
2	.61	.54	.57	.31	.43	.72	1.26
3	1.09	.95	1.02	.45	.72	1.45	1.42
4	1.66	1.41	1.53	.51	.94	2.39	1.56
5	2.24	1.89	2.06	.53	1.07	3.46	1.68
6	2.91	2.42	2.67	.61	1.26	4.72	1.77
7	3.66	3.00	3.33	.66	1.45	6.18	1.86
8	4.38	3.58	3.98	.65	1.62	7.79	1.96

*Feed conversion = Pounds of feed consumed divided by the live weight
Based on information from the New England Poultry Management and Business Analysis Manual, Bulletin 566 (Revised).

TABLE 15

WATER CONSUMPTION OF BROILERS

	Gallons Water per 100 birds	
Age in Weeks	**Weekly**	**Daily**
0-1	7.0	1.0
1-2	10.3	1.5
2-3	17.3	2.5
3-4	22.5	3.2
4-5	25.7	3.7
5-6	30.2	4.3
6-7	34.8	5.0
7-8	38.8	5.5

Source: 1972-73 *New England Poultry Management and Business Management Manual — Bulletin 566.*

TABLE 16

ESTIMATED PER BIRD COSTS OF BROILER PRODUCTION[*]

	Cents
Chick cost	16-20
Brooding	02-03
Feed (7.79 lbs. at .08)	62-62
Litter & Misc.	02-04

Total Per Bird: 82-89 cents

*Interest on average investment, labor and overhead (depreciation, insurance and taxes) are included in the costs of producing commercial broilers but frequently omitted in small flock situations. The approximate costs of these items are: interest — 2¢, labor — 2¢, and overhead — 6¢.

Housing And Equipment Needs

Meat-type birds can be grown either in confinement or on range. However, most broilers are grown in confinement. For at least the first several weeks they have to be confined to the brooder house. Since the males may be dressed as early as the sixth week it usually doesn't pay to move them to range. The best growth results are obtained by rearing in confinement.

Roasters and capons are sometimes grown on range or yarded. This is satisfactory where housing is scarce, but growth and finish may be sacrificed by so doing. Meat birds usually do better when confined and receiving a full feed of a specialized ration designed for rapid growth and efficient conversion.

floor space

Meat birds should receive at least ½ square foot per bird until the chicks are two weeks old and one square foot per bird between the second and tenth week. Heavy meat birds such as capons and roasters from 10 to 20 weeks of age should have 2-3 square feet of floor space per bird. When meat type birds are kept over 20 weeks they should have a minimum of 3 square feet per bird and more if available.

feed and water space

The chicks may be fed in box lids for the first 5 to 10 days, or chick-size feeders may be used at the rate of 1″ per chick. At 3 to 6 weeks, they should have 2″ of feeder space per chick and at 7-12 weeks, 3″. Birds to be carried on as roasters should be provided with 4″ of feeder space per bird for the remainder of the growing period.

Water should be provided at the rate of 20″ of hopper space or two 1 gallon waterers per 100 chicks up to 3 weeks of age. From the third week they should have a 4 or 5 foot trough per 100 chicks or two large fountain type waterers.

The feed hopper should not be filled more than ½ full to avoid feed waste. The feeders should be adjusted so that the lip of the hopper is at a level with the bird's back. Adjustable hanging tube feeders are excellent for growing birds. Three tube-type feeders will normally accommodate 100 birds.

housing requirements

General housing requirements — the requirements for cleanliness, sanitation, litter, the arrangement of feeding and watering and brooding equipment — are all much the same as those described for brooding replacement pullets in *Chapter 4*. Roosts are not recommended for brooding meat birds. Roosting can cause such problems as crooked breast bones and breast blisters.

Lighting meat birds is somewhat different than lighting replacement birds. Enough light should be provided the broiler chick to enable it to move about and see to eat and drink. Actually, it is desirable to keep activity at a reduced level for the most efficient feed utilization. The intensity of illumination at the level of the chick should be about ½ of a foot candle. It is sometimes difficult to keep this level of lighting in housing with windows. Birds that receive a higher intensity are more prone to cannibalism, increased activity and possible piling and smothering. One bulb watt per 8 square feet of floor space will normally provide ½ foot candle intensity. This is when the bulb has a reflector and is 7 to 8 feet above the floor.

Many broiler producers use a continuous-light program. This may be somewhat hazardous. In the event of a power failure, the birds may panic and pile. A light-day of 14 hours is usually sufficient. If the birds are given enough light to enable them to consume adequate feed and water, it is not usually necessary to provide more than 14 hours. When the weather is exceptionally hot, it may be an advantage to use a 16 hour light day to provide additional feeding time during the cooler hours of the morning or evening.

management systems for meat birds

Although some research work is being done to develop cage equipment for broilers and other meat birds, the recommended management system is still the conventional floor litter system. Cage growing offers several advantages but, thus far, the disadvantages outweigh these advantages. More breast blisters, leg and foot problems, crooked keels, poor growth and feed efficiency are problems which may result with cage management systems of growing meat birds.

The windowed house or controlled environment houses are used for meat birds depending upon the climate, the size of flock and other factors. Ventilation and ventilation systems are the same as those for other types of growing houses. In most areas concrete floors are required for proper cleaning and sanitation and to avoid disease and parasite problems. The cleaning and sanitation requirements are the same for meat birds as for other growing stock.

disease control

Disease prevention is vitally important to a broiler growing program. Unlike birds that are grown for egg production, or meat birds such as the roasters or capons, broilers have only an 8 week growing period. If the birds become sick in the middle of the growing period, the time in which the disease can be treated and brought under control is relatively limited, and the overall flock results can be severely affected. It is, therefore, important that a disease control program for broilers be one of prevention rather than treatment.

Vaccination recommendations for meat birds vary with the area. It is probably well to plan to have the chicks vaccinated for Marek's disease, especially if the disease has been prevalent in your area. Marek's vaccine is usually administered at the hatchery. Some broiler growers do not vaccinate for other diseases, because they practice isolation and sanitation practices and feel that these programs will prevent most disease outbreaks.

Before establishing a vaccination program for meat birds, be sure to check with poultry pathologists or other people who are knowledgeable about the possible disease exposure situation in the area. When broilers or other meat birds are vaccinated for other than Marek's disease, it is normally Newcastle disease or infectious bronchitis.

It is very important to prevent outbreaks of coccidiosis in meat birds particularly during the broiler stage. Outbreaks of coccidiosis normally occur early in the chick's life and can damage the intestinal lining. This prevents the normal absorption of food materials in the intestinal tract. Outbreaks of coccidiosis can cause poor growth and feed conversion, or — in severe cases — mortality and morbidity.

Drugs called coccidiostats are normally used in the feeds to either reduce the infection of coccidiosis or to completely suppress it. When pullets are grown for a floor management system of housing layers, the starter feeds and grower feeds contain low levels of a coccidiostat which permits the birds to have a low level of infection and build up immunity to the disease. Birds that are grown on the floor and are to be housed in cages — where they won't be exposed to fecal material and contained sporulated oocysts — are given a type of coccidiostat, or dosage of coccidiostat, in the diet to prevent infection altogether.

Since the growing period for broilers is so short, and so much emphasis is placed on fast efficient growth, the usual procedure is to control coccidiosis during the growing period. Better growth results are obtained if infection is prevented. Most roaster and capon growers use a coccidiostat throughout the growing period.

Even though the birds are on a preventive level of coccidiostat, at times there are outbreaks of the disease due to excessively wet litter or other factors. The method of treating these outbreaks is covered in the section on flock health, *Chapter 13.*

feeding programs for meat birds

The goal of the feeding program for meat birds is to get as much growth as quickly as possible. This is unlike feeding replacement pullets where we are interested in delaying sexual maturity. Meat birds must have access to full feed at all times, and they should be encouraged to eat as much as possible, with as little waste as possible. A number of small flock owners permit capons and roasters to yard or range. Though overall performance may be somewhat less, excellent birds can be grown this way. The cost may be somewhat less also, because the chickens forage for a part of their food intake.

It is important to feed meat birds a ration that is designed to give rapid growth and good feed conversion. They should be started on a broiler feeding program. The complete feeding program includes two or possibly three different rations. Some companies recommend a program which begins with a broiler starter the first five weeks. The starter contains about 22% protein and a coccidiostat. This feed is replaced with an 18% broiler finisher from 5 weeks up to 5 days prior to marketing. At this time they would be shifted to a final feed of 17%, but containing no coccidiostat. Where this feeding program is used, it should be stressed that the birds should be checked very carefully for coccidiosis if they are to be held beyond the 5 day period. It may become necessary to treat for coccidiosis if the birds are held more than 5 days prior to dressing or marketing.

Where birds are to be held over as light roasters for 4 to 5 weeks beyond the broiler period, they can be fed a 20% broiler finisher diet. This may be fed up to within 2 to 5 days prior to dressing or marketing, depending upon the type of coccidiostat used. Some coccidiostats must be withdrawn several days prior to slaughter.

Where birds are fed as capons or heavy roasters, they should be fed a lower protein and higher energy diet from the ninth or tenth week to marketing. The protein level should be approximately 16%. Some feed company programs call for feeding a roaster finisher from the sixth week to 2 days before marketing, at which time a feed containing no coccidiostat is used for the remainder of the period.

Capon Production

Male chickens, sometimes castrated or caponized to make them fatten more readily, are called capons. Capons are not only more tender than cockerels but bring a much better market price. Capons are normally grown to be marketed during the Thanksgiving-Christmas holiday seasons at which time they bring premium prices.

Up to possibly 20 weeks of age the capon and the cockerel weigh about the same. Afterwards, the capons gain weight more rapidly than cockerels. When cockerels show spur and comb development and are called stags, their flesh toughens. Toughness of the flesh does not occur in capons. When cockerels are a year old, and are classed as old cocks, they bring a very low price, and their flesh is very tough. On the other hand, capons have very tender flesh, and if kept beyond the age when they are normally marketed, or dressed-off, the flesh remains tender.

Unlike the cockerel, the capon has a very quiet disposition, is very docile and seldom crows. After caponization, the comb and wattles cease growing, the head looks pale and small, and the hackle, tail and saddle feathers grow to be unusually long.

selecting a breed for caponizing

Broiler stocks with white plumage and yellow skin, are preferred for meat bird production. White Plymouth Rocks, New Hampshires and crosses of these breeds or those crosses involving White Cornish and White Rock Crosses are generally used. Caponizing small birds, such as Leghorns, or the small representatives of the American breeds, just does not pay.

when to caponize

Since capons are in the biggest demand during the Thanksgiving to March period and take 24-25 weeks to grow and finish properly, the best time to caponize is in late spring or early fall.

Caponizing used to be done at 5-6 weeks of age. With our faster growing broiler strains of birds, it is now desirable to caponize at approximately 4-5 weeks of age before they become too large, when the operation is more difficult. Some hatcheries or dealers specialize in selling 4-week old capons that have been caponized as early as 10 days. It is possible to caponize older birds up to 8 weeks; it is best, however, to do it earlier.

the caponizing operation

Feed and water should be withheld from the birds for 24 hours before the operation. The bird's intestines should be almost empty to prevent them from obstructing the operation. When the birds are removed from feed and water, they should be kept in wire or slat-bottom coops to prevent them from eating litter or other materials.

Good light is mandatory during the operation. Direct sunlight is best; a strong electric light with a reflector can, however, serve indoors.

A barrel or box can be used as an operating table. When a large number of birds are to be caponized, a table of convenient height is recommended.

During the operation, the birds can be restrained by a second operator who holds the bird outstretched on its side. A caponizer can do the operation single-handedly by restraining the birds with weighted cords tied to the legs and wings. These cords should be about 2 feet long and weighted with 1 pound weights. The cockerel's legs are securely tied by the use of a half-hitch in one cord. Both wings are held together near the shoulder joint with a half hitch in the other cord. The weights are hung over the edge of the operating table, and the bird is stretched out. The cords should be so arranged that it is easy to adjust the bird and turn it over without disturbing the weights.

methods

The operation can be done with one incision or two incisions. Most operators find it easier to remove the upper, or nearer testicle, then turn the bird over, make a second incision on the other side of the body for removing the other testicle. If both testicles are removed through the one incision, it is best to remove the lower one first; otherwise, bleeding from the upper may obscure the lower.

It is very important not to rupture the artery which runs just behind the testicles. If the artery is ruptured the chicken will bleed to death in seconds. Usually if feed and water have been withheld, the artery presents no problem.

The instruments which are needed for the operation are: a sharp knife, a rib spreader, a sharp pointed hood and the testicle removers, or forceps *(Figure 18)*. The operation should be performed as quickly as possible to reduce stress on the bird.

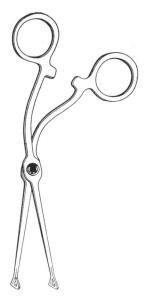
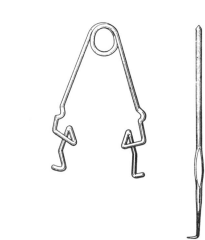

(Figure 18. Forceps, Knife, Rib Spreader, Sharp Hook.)

Moisten and remove the feathers from a small area over the last two ribs, just in front of the thigh; with one hand, slide the skin flesh down toward the thigh and hold it in that position. This will prevent cutting into the muscle and causing excessive bleeding. The cut is made between the last two ribs *(Figure 19)*. The incision should be lengthened in each direction until it is about ¾" to 1" long.

The spreader is then inserted into the incision to spring the ribs apart *(Figure 20)*. The intestines will be visible beneath a thin membrane. This membrane should be torn open with the hook. This procedure will expose the upper testicle. It is usually yellow in color but may be dark colored. It is about the size and shape of a navy bean and is located up close to the back bone and just below the front end of the kidney. If the intestines are empty, once they are pushed aside, it is quite easy to see the lower testicle. It is in the same position as the upper one, but on the other side of the backbone.

(Figure 19.)   (Figure 20.)

The forceps are used to grasp the testicle and it is important that the artery be avoided at this point. The entire testicle must be removed with a slow twisting motion, tearing it away from the spermatic cord to which it is attached. Unless both testicles are removed from the same incision, the bird is turned over, and the same operation is performed on the other side. It is not necessary to stitch the incision, since when the skin and flesh are released, they will slide back over the incision and cover it.

slips

Frequently, when the caponizing operation is performed for the first time, a number of birds will become what are known as slips. A "slip" is neither a cockerel nor a capon. A slip develops when a part of the testicle is not removed during the operation. This small piece of testicle often grows to a considerable size. Slips have the same restless disposition as cockerels and they grow and fatten little, or no better, than a cockerel. Thus slips are not as highly valued for meat purposes and do not bring as good a price if sold. Inexperienced operators may expect up to 50% slips; experienced operators can expect about 5% or less.

losses in caponizing

Occasionally, even the best operators will kill some birds during the caponizing operation. However, this loss seldom exceeds 5%. Inexperienced caponizer frequently kill several birds, but gradually losses are reduced as they gain experience. Any birds that are killed during the operation may be dressed and eaten.

Speed develops with experience. An experienced caponizer can do up to four birds per minute. It is wise for the beginner to practice, at first, on slaughtered birds to acquire some skill before attempting to caponize a live bird. Caponizing is somewhat of a stress on the birds, so some commercial caponizers inject the birds with an antibiotic solution just prior to the operation. Some capon growers administer the antibiotic solution in the drinking water for 2 or 3 days, beginning 4 or 5 days prior to the operation. Antibiotics should not be given by way of the drinking water within two days of the operation. Some antibiotics tend to cause the intestines to balloon, thus increasing the possibility of damage to the intestine and making it more difficult to spot the testicle.

care and feeding of caponized birds

Caponized birds should be separated from other chickens and kept separated during the growing period. Air puffs or wind puffs may develop after the operation. Wind puffs are caused by air which gathers under the skin. This air should be released by pricking the skin with a needle or knife and pressing it out. Usually within ten days after the operation, the incision is fully healed.

Breast blisters are often a problem with capons. They are likely to appear when the capons are about half-grown and the incidence may increase as the birds become heavier. The birds may develop breast blisters from roosts, especially those with narrow perches or from resting on board, concrete, or wire floors. Litter should not be permitted to become caked or excessively wet but should be kept soft, loose and dry. Roasters or capons should not be permitted to roost.

During the growing period, the birds should be supplied with a growing feed which contains approximately a 17% protein. The growing stage is very important because during this period leg problems may occur. Capons, if pushed too hard for rapid growth during the early growing period, may have insufficient bone development to hold up the body weight. Most cockerels that are to be used for capons are actually broiler males. This stock has been selected for rapid growth and males should reach 3½ to 4 pounds in body weight at 8 weeks of age. They are normally marketed at these weights. When we use these birds for capon production it is expected that they will grow to about 9 or 10 pounds live weight in approximately 20 weeks. Some of the birds will not have strong enough legs to support these heavier weights if growth is too rapid early in the growing period. It is, therefore, well to keep this in mind when selecting a feeding program for capons. Some growers restrict feed, and thus the growth rate, from about 6-10 weeks of age to prevent this problem in light roasters. Feeds especially formulated for capons take this problem into consideration. Broiler rations, however, are high energy feeds that are designed for maximum growth and may create the leg problems if fed throughout the growing period. For heavy roasters, it is suggested that a lower energy diet be used from approximately 8 to 13 weeks.

Leg weaknesses also contribute to the breast blister problem. This problem can be very serious, because it detracts from the dressed appearance of the birds. Birds that do have a weak leg problem exhibit a tendency to sit down much of the time. The keel bone is constantly in contact with the litter and the incidence of blisters is increased, especially if litter conditions are poor. Other factors that affect the incidence of breast blisters include the type of equipment used, overcrowding, genetics, loading and handling methods and others. Birds can receive bruises and skin irritations by bumping into equipment or by being pushed and shoved or mishandled.

Capons may be grown in confinement with a yard provided or on range. Downgrading often seems higher in the case of confinement reared birds. The reason for this is that confined birds are usually pushed to higher weights more quickly. There is also a greater possibility of crowding and injuries when birds are confined.

Range birds, however, seem to have higher mortality. A lot of this is probably due to predator losses. Birds on range should be provided with shelters. Often, shelters with wire floors and perches are used. These may cause breast blisters. Wire floors and perches should be covered, and preferably litter used on the floor of the shelters. Shade is important for birds on range. This is especially true during the hot weather seasons, or in warm weather climates. The range should be clean and provide a short succulent forage to be most valuable. If good range is provided, feed costs can be substantially reduced. Birds on range should be provided with range feeders that prevent the feed from being blown away or getting wet and subsequently moldy. Adequate feeder and watering space is important regardless of whether the birds are reared on range or in confinement. Approximately 4″ of feeder space and 1″ of watering space per bird is required.

When finishing capons, the birds should be pushed for maximum gains and efficiency after they reach an age of 13 weeks. The protein levels of the feed can be reduced to approximately 15 or 16%, and the energy level increased to obtain maximum yet efficient weights, as well as to get the birds in good market condition in terms of fat cover, fleshing, and pigmentation.

chemical caponizing

Dienesterol Diacetate (Lipamone) may be used to produce chemical capons. Lipamone is added to the finisher diet and is fed to the birds for the last 3 or 4 weeks prior to slaughter. This program is often used for birds that are to be slaughtered as light roasters at 12-14 weeks. Lipamone-fed birds seem to have more fat beneath the skin and, therefore, the bird is more tender and succulent and has an improved flavor.

Another program which we have used in Vermont for a number of years makes use of a product which is relatively easy to administer, and yields essentially the same end results as surgical caponizing. A steroid paste (Esmopal) is administered with a special gun which meters the correct amount of material through a needle. The implant is injected just under the skin alongside the comb. This material is manufactured by the Mattox and Moore Company, Indianapolis, Indiana, and comes in a 50-dose cartridge. Birds so injected take on much the same appearance as surgical capons. The comb and wattles become pale and small, and the birds become quiet and somewhat more docile than untreated birds.

when to implant

Once the birds are implanted, a minimum of 6 weeks time must have elapsed before they can be slaughtered for human consumption. Birds should be 5 weeks of age before implanting. Those injected at 5 weeks of age must be at least 11 weeks of age when slaughtered. It is absolutely necessary to plan the marketing or slaughtering program. At the end of 6 weeks following the implant, birds must be slaughtered or re-implanted, because the steroid material is completely absorbed by that time, and they will begin again to appear and act like cockerels.

In trials conducted at the University of Vermont, Bioresearch Center, birds were implanted at 5 weeks, again at 11 weeks and slaughtered at 17 weeks. Live weights of 9½-10½ pounds were reached. If heavier market weights are desired the birds could, for example, receive the first implant at 7 weeks, a second at 13 weeks and be slaughtered at 19 weeks. With this program some birds can be slaughtered as light roasters at 11 or 12 weeks of age or they can be reimplanted and carried on to heavier weights. Feed consumed is approximately 38 pounds per bird at 17 weeks. Implanting costs approximately 12 cents per bird on the 17-week program. Estimated costs of production of roasters and capons is shown in *Table 17*. Care and feeding of the chemical roasters is the same as for capons.

TABLE 17

APPROXIMATE PER BIRD COSTS OF PRODUCING CAPONS AND ROASTERS*

Item	Cost
Chicks	$ .16 — $ .20
Brooding	.02 — .03
Feed, 38 lbs. at $.08-.09	3.04 — 3.42
Litter and Miscellaneous	.04 — .05
Cost of Esmopal or Surgical Caponizing	.12 — .20
Total:	**$3.98 — $3.90**

*Interest on average investment, labor and overhead (depreciation, insurance and taxes) are included in the costs of producing commercial roasters but frequently overlooked when grown in small flocks. Estimated costs of these items are: interest — 7¢, labor — 12¢, and overhead — 17¢.

9

TURKEY PRODUCTION

Before you launch into the production of turkeys, even on a small scale, you should be aware of the costs of production. The turkey poult costs substantially more than does a chick and, of course, turkeys consume considerably more feed than do growing chicks. Thus the cost of producing turkeys is considerably higher. *Table 18* presents the costs involved in raising a small flock of turkeys.

Getting Started

The turkey enterprise can be started in one of several ways. You can purchase hatching eggs and incubate them, or purchase day-old poults. One may also buy breeding stock, but this is a rather expensive way to begin. If hatching eggs are purchased, it requires equipment for incubating; if breeding stock is purchased it is not only expensive to buy the birds but also to maintain them. Normally, most small flocks of turkeys are purchased as poults from a hatchery.

There are several varieties of turkeys, but those that are most important commercially are the Broad Breasted Bronze, the Broad

TABLE 18

ESTIMATED PER BIRD COSTS OF RAISING HEAVY ROASTER TURKEYS

Item of Cost	Amount
Poult	$ .75 — $1.00
Feed (75.5 lbs.)	6.04 — 7.55
Brooding, Electricity, Litter, Misc.	.15 — .20
	Total: $6.94 — $8.75

Assumptions: Small flock, their feed costs are relatively high, poult costs are usually high, all feed purchased, no labor, overhead, or interest costs included.

Breasted White, and the Beltsville Small White. Until quite recently the Bronze was the most popular variety.

The Bronze has good growth rate, confirmation, feed conversion and about all of the qualities demanded by the turkey industry. Its basic plumage color is black, and it has dark colored pin feathers, a distinct disadvantage that has led to its replacement by the Broad Breasted White.

The Broad Breasted White was developed in the early 1950's from crosses of the Broad Breasted Bronze and the White Holland variety. In some areas of the country it is now difficult to find the Broad Breasted Bronze, because the current demand is so great for the Whites.

The Small White looks very much like the Large White from the standpoint of color and body conformation. It is a much smaller turkey which was developed at the Beltsville Agricultural Research Center in Maryland. Its body weight has tended to increase in recent years through selection by breeders.

Properly managed and fed the Beltsvilles reach good market condition in 15 to 16 weeks and make excellent turkey broilers. They also make good medium roasters if held to 21 to 24 weeks. They are not as efficient converters of feed as the Large Whites but usually cost somewhat less as poults. Their main disadvantage is the difference in feed required to reach the same weights as the large turkeys.

When buying poults you should select a strain that is known to yield good results. As is the case with chickens, the poults should originate from sources that are U.S. pullorum-typhoid clean, and preferably from breeder flocks having no history of sinusitis, or air-sac infection. Consult your state poultry specialist, your county Agricultural Extension Agent, or some other knowledgeable person for advice on sources of good turkey poults in your area.

Place your order for poults well in advance of the delivery date, so as to be sure to get the quality of stock you want. Birds should be ordered to arrive so as to permit 24-28 weeks for growing the large breeds and 18-22 weeks for smaller breeds. It is sometimes difficult to obtain small lots of turkeys delivered from the hatchery. However, it may be possible to pick the turkeys up at the hatchery or, perhaps, have a nearby producer order a few extra poults for you. Poults that are shipped express are subject to chilling and overheating. Buy the poults as close to home as possible.

housing requirements

Small flocks of turkeys are normally started in the warm months of the year so housing doesn't have to be fancy. However, the brooder house should be a well-constructed building that is easily ventilated. If a small building is not available a pen can be built within a large building. The pen should have good floors that can be readily cleaned. The insulation required will depend upon the climatic conditions as well as the time of year the poults are to be brooded. A well-insulated building can save energy and brooding costs. Young turkeys must be kept warm and dry, so a well insulated and ventilated house is important in cool climates, particularly with large flocks.

The brooder house should have windows which slide down or tilt in from the top. These are best for ventilation. The windows should be located in the front and back of the house. Usually one square foot of window area to each 10 square feet of floor area is adequate.

Confinement rearing of turkeys is most frequently used by the small grower. Where predators or adverse weather conditions are likely, or where there is limited range area, confinement rearing is preferred. Some producers yard their birds within fenced-in areas using the brooder house as the shelter.

Sun porches were once very popular with turkey growers. Some producers still use them. The porches are attached to the brooder house or shelter. The floor of the porch is made either of slats or wire. They are frequently elevated to provide a space underneath for the accumulation of droppings and an easy access for cleaning. Porches are fenced in on the top and sides. Coarse mesh wire is used on the top to avoid snow loading. Porches have several advantages in that they help to acclimate the birds to the weather conditions prior to going on range. This helps to reduce stampeding on the range during wind or rain storms. The porches also acclimate them to changes in temperature to which they are exposed on range.

The heavy breed turkeys today are not grown on wire or slated porches as much. Breast blisters, foot and leg problems tend to develop particularly on wire. Light type turkeys may still be grown on porches. A modification of the porch approach is the use of paved yards or yards with a gravel or stone surface.

Large turkey production units sometimes use windowless, controlled environment houses. These buildings are thoroughly insulated to provide maximum efficiency in the use of heat during brooding and to provide comfortable well-ventilated conditions for the birds.

Ventilation systems for turkey housing are designed essentially the same as for chicken houses. Many of them use fan ventilation with the same type of inlet but a larger ventilating capacity. Fan systems are designed to move from ¾ to 1 cubic foot per minute at ⅛ inch static pressure per pound of turkey expected at maturity. If you plan to go into the turkey business on a commercial scale, contact your State Extension Poultry Specialist for help in designing the facilities.

preparing the brooder house

The brooder house should be cleaned between broods. This means completely removing all the old litter and thoroughly washing the floors, sidewalls, ceilings and equipment. The building and equipment should be disinfected with a material such as cresilic acid. Some of these materials can cause foot burn or eye injury. To prevent disinfectant injury to the poults, make sure that the house has an opportunity to dry for about 2 weeks prior to the time the poults are put in the house.

After the building is thoroughly dry, 3" to 4" of litter should be put down on the floor. Some of the common litter materials which are suitable for use are wood shavings, sugarcane, ground corn cobs, peatmoss or vermiculite. The litter may be covered or uncovered. Some producers cover the litter with paper to prevent litter-eating for the first week. If the litter is covered, a rough paper should be used. A slippery surface can cause leg weakness and crooked feet. The litter should be evenly distributed over the floor and free of mold and dust. Very coarse litter material can also contribute to leg disorders.

In small brooder houses, corners should be rounded with small mesh wire to prevent piling. Turkeys may pile when frightened or when the house is drafty or floor temperatures low.

floor space

Use one square foot of floor space per poult up to 8 weeks of age. From 8-12 weeks the floor space should be increased to 2 square feet per poult, and from 12-16 weeks two and one half square feet should be the minimum allowance. Mixed sexes which are to be kept in confinement during the entire growing period should receive 4 square feet per bird. If the flock is all toms, they should receive 5 square feet or, if all hens, 3 square feet is adequate. For light type turkeys, or for turkeys housed in controlled environment housing, the floor space requirements may be somewhat less than the above recommendations.

brooders

Several types of brooders are suitable for brooding poults. Gas or electric are probably the best suited for small flock situations. If hover type brooders are used, allow approximately 12-13 square inches of hover space per poult started. A brooder that is rated for 250 chicks will be adequate for only about 125 turkeys. When hover type brooding is used, a 7½ watt light bulb should be used to attract the poults to the heat source. Each brooder should be equipped with a thermometer which can be easily read. Infrared brooders are satisfactory for small turkey flocks.

If infrared bulbs are used for brooding, provide 2 or 3,250 watt bulbs per 100 poults. Infrared lamps should be hung about 18 inches from the surface of the litter initially. They can be raised 2 inches per week until 24 inches above the litter. The brooder ring or guard should be 8-10 feet in diameter. Care should be taken to avoid splashing water on infrared bulbs. This may cause them to break unless they are the Pyrex-type. Ideally, the temperature outside the heat source or the hover area should be approximately 70° F. to provide maximum comfort for the poults.

brooder guards

Brooder guards should be used to confine the birds to the heat source and to prevent drafts on the poults.

Management of the brooder guard varies, depending upon the design of the house and the climatic or seasonal conditions. Where non-insulated housing is used during fairly cool weather, an 18 inch brooder guard for each stove is recommended. For warm weather brooding a 12 inch brooder guard is satisfactory. The brooder guard should be located 2-3 feet from the edge of the hover and be gradually moved out to a distance of 3 to 4 feet. After 7 to 10 days the brooder guard can be removed. At this time the poults are normally allowed free access to the brooder house. If poults are brooded early in the season, or in cold climates, and the brooder house is quite large, it may be necessary to partition off part of the house at the start and enlarge the area as necessary.

feeding equipment

The first feed for the poults can be supplied to them on new egg filler flats, on chick box lids, or in small chick feeders. Turkeys sometimes have a visual problem and have difficulty finding the feed and water. When this occurs there may be starve-outs or dehydrated poults. Bright colored marbles placed on top of the feed or in the water containers will often attract the poults to the feed and water. Oatmeal or fine granite grit sprinkled very lightly over the feed, once or twice a day for the first 3 days, may also help to get them eating. Unless the litter is covered with a paper, don't fill the feeder so full that it will overflow. This can lead to the practice of eating litter. For poults from 7 days to 3 weeks, use small feeders at the rate of 2 inches of feeder space per bird. From 3 weeks to market, the poults should be furnished with large feeders about 4 inches deep at the rate of 3 inches of feeder space per bird. Hanging tube-type feeders can also be utilized. The amount of tube type feeding space can be determined by multiplying the diameter of the feeder pan by 3 1/7 to obtain the number of inches of feeder space available. In figuring feeder space remember to multiply the hopper length by 2 if the poults are able to use both sides of the feeder. Thus a 4 foot trough feeder actually provides 8 linear feet of feeder space.

water space

Poults can be started on either glass fountain type waterers or automatic waterers. From 1 day to 3 weeks the poults should have access to 3, one or 2 gallon fountains per 100 poults. From 3 weeks to market they should have 2, 3 or 5 gallon fountains per 100 poults, or one 4 foot automatic waterer.

Changes of equipment, both feeders and waterers, should be made gradually so as not to discourage feed or water consumption.

brooding the poults

When the poults arrive the brooder house should be completely ready for them. The feeding equipment should be arranged and the brooder guards set up (See *Figure 10)*. The brooders should have been in operation for approximately 24 hours to get them regulated and to get the building warm.

If hover type brooders are used, the temperature should be 100° F. for white birds, 95° F. for Bronze. This temperature reading is taken under the hover approximately 7 inches from the outer edge and 2 inches above the litter, or at the height of the poults' backs. Be sure to check the accuracy of the thermometer you are using before the poults arrive. The hover temperature should be reduced 5° weekly until it registers 70 or 75° F. or equals the outside temperature. If the weather is warm during the brooding period, heat may be shut down during the day after the first week. Heat during the evening hours will be required for a longer period. Normally, little or no heat is required after the sixth week, depending upon the time of year, the weather conditions and the housing. Watch the poults as a guide when checking or adjusting temperatures *(Figure 11)*.

Important: Feed and water the poults as soon as possible after hatching. As each poult is taken from the box, dip its beak in water and then in the feed. This will help to get it started on feed and water.

lighting

High light intensity should be provided the poults for the first 2 weeks of brooding in all types of houses to prevent starve-outs. Ten to 15 foot candles of light should be used day and night. This will require 200 watt bulbs spaced 10 feet on centers. A small 7½ to 15 watt attraction light bulb should be installed under each brooder hover. After the first two weeks in window houses, use dim lights at approximately ½ foot candle during the night hours. The dim lights will help to prevent, or discourage, piling and stampeding.

roosts

Roosts are seldom used for brooding turkeys though they may help to prevent piling at night. Roosts do not normally cause breast blister problems with turkeys. If roosts are used in the brooder house they may be of the step ladder type, allowing about 3 linear inches of perch space per bird. The roosts perches are made of 2 inch round poles or 2″ x 2″ or 2″ x 3″ material, with the lowest perch located about 12 inches above the litter. Each succeeding perch is 4-6 inches higher. The perches should be beveled or rounded on the edges to prevent injury to the breasts. Each bird should have 6 linear inches of roosting space by the end of the brooding period. The bottom and the ends of the roosts should be screened so as to prevent the poults from gaining access to the droppings. Birds usually begin to use the roosts at about 4-5 weeks of age.

Usually where birds are grown in complete confinement roosts are not used. Since the birds bed down on the floor, litter conditions should be kept in good condition to prevent breast blisters, soiled and matted feathers, off-colored skin blemishes on the breast. Good litter conditions also help to maintain sanitary conditions and prevent disease problems.

Roosts which are used on range should be constructed of 2″ poles or 2″ x 4″ material laid flat with rounded edges. The roosts perches should be spaced 24″ apart and located approximately 3 feet off the gound. If placed in a house or shelter, the roosts may be slanted to conserve space. Where an outdoor roosting rack is constructed, all roosts perches should be built on the same level. This type of roost should be built of fairly heavy material to prevent breaking when the weight of the birds is concentrated in a small area. Ten to 15 inches of perch space, per bird, is required for the large type birds and 10 to 12 inches for small type turkeys.

feeding

One of the soundest pieces of advice that can be given to a turkey grower is to select a good brand of feed and follow the manufacturers recommendations for feeding the birds.

Two basic feeding programs for growing turkeys are available. One is the all mash system and the second is a protein supplement plus grain system. Nutrient requirements of turkey poults vary with the age. As the birds become older the protein, vitamin and mineral requirements decrease and the energy requirements increase.

An insoluble grit such as granite grit should be fed the first 8-10 weeks. Whenever grains are included in the diet or if the birds are on range they should receive an insoluble grit to enable them to grind and utilize the grains and fibrous materials.

Recommendations for feeding, and the number of feeds offered, vary considerably between feed companies. One of the simpler types of feeding programs recommends a 28% starting diet with a change to 21% growing diet and finishing with a 16% protein diet. Other commercial concerns offer and recommend 5 or 6 different diets during the growing period. Again, buy feed from a reputable feed company and follow their recommendations. Under most circumstances the diets should include a coccidiostat and, in most instances, there should be a blackhead disease preventative drug incorporated in the feed.

Feed and water should be kept before the birds continuously. Pelleted mash can be fed after the first 4 weeks. Most growers feed a nutritionally complete starting mash. However, green feed can be utilized for small flocks where labor requirements are of little concern. Tender alfalfa, white Dutch clover, young tender grass or green grain sprouts all chopped into short lengths and fed once or twice daily are good for the poults. Wilted or dry roughage feeds should not be permitted to remain before the poults. These can cause impacted or pendulous crops.

Turkey starter diets can be purchased ready-mixed. They can be home or custom-mixed according to recommended formulas, or a concentrate can be used and mixed with ingredients such as ground corn and soybean meal. For the most part, small producers will find it best to feed a ready-mixed complete starter diet.

Growing diets, fed to the poults from 8 weeks to maturity, may be in mash or pellet form. They may include either loose or pelleted mash plus whole or cracked grain — sometimes referred to as scratch grains. A commercial concentrate may also be purchased and combined with ground grain or with soybean meal and ground corn in the proportions recommended by the manufacturer. Once again the small grower will find it advantageous to use a complete ready-mixed mash when the birds are reared in confinement. If the birds are on range, a complete feed, preferably

TABLE 19

GROWTH RATE AND FEED CONSUMPTION FOR HEAVY AND LIGHT ROASTER TURKEYS AND BROILER FRYERS

| | BROAD BREASTED BRONZE AND BROAD BREASTED LARGE WHITES | | | | SMALL WHITES OR BROILER-FRYER-ROASTERS | | | |
| | Live Weight Pounds | | Total Cumulative Feed Required Pounds | | Live Weight Pounds | | Total Cumulative Feed Required Pounds | |
Age Weeks	Toms	Hens	Toms	Hens	Toms	Hens	Toms	Hens
2	.55	.53	.6	.53	.5	.45	.6	.55
4	1.4	1.2	2.1	1.8	1.5	1.3	2.25	2.0
6	2.5	2.0	4.8	3.8	2.9	2.3	4.5	3.8
8	4.0	3.5	8.8	7.7	4.6	3.6	8.0	6.7
10	5.4	5.2	16.2	14.1	6.9	5.3	13.0	10.6
12	8.2	7.0	25.0	21.0	9.4	7.1	20.2	15.7
14	11.0	9.0	34.3	28.0	11.8	8.5	28.0	21.7
16	14.0	11.0	43.8	34.2	14.2	9.8	36.1	28.2
18	16.9	13.0	52.9	40.3	16.4	10.9	45.2	34.9
20	19.7	14.8	62.5	49.0	18.5	11.8	56.0	41.7
22	22.4	16.3	74.0	57.7				
24	25.2	17.3	87.3	64.0				
26	28.0		103.0					
28	30.6		120.0					
Total	30.6	17.3	120.0	64.0	18.5	11.8	56.0	41.7

Based on material by M. L. Scott, *Turkey World*, 1970.

in pellet form, supplemented with good range, grains and insoluble grit makes a sound feeding program. Growth rate and feed consumption information for heavy and light roaster turkeys is presented in *Table 19*. Water consumption information is presented in *Table 20*.

TABLE 20

DAILY WATER CONSUMPTION OF TURKEYS GALLONS PER 100 BIRDS

Age in Weeks	Large Turkeys	Small Turkeys
1	1	.7
2	2	1.5
3	3	2.5
4	4	3.5
5	5	4.0
6	6	4.5
7	7.5	5.0
8	9.5	5.5
9	11.0	6.5
10	12.5	7.0
11	14.0	7.5
12	15.0	8.0
13	16.0	9.5
14	16.5*	10.5
15	17.0*	12.0
16	16.5*	13.0
17	16.5*	—
18	16.5*	—
19	16.5*	—
20	16.5*	—

Dr. Salisbury Laboratories
*Consumption will vary from 15 weeks to maturity from 14.0 to 19.0 gallons per 100 birds per day depending upon the temperatures.

range-rearing

It is possible to reduce the cost of rearing turkeys by putting them on range. This is especially true if the diet can be supplemented with home grown grains. Turkeys are good foragers and if good green feed is available on the range this will help to supplement the diet and reduce the cost of the feeding program.

Range-rearing is not without its problems. Losses are possible from soil born diseases, insects, thieves, adverse weather conditions, and predators. Because of these several factors confinement rearing, which tends to minimize these losses, has quite rapidly replaced range rearing in some areas.

Portable range shelters which can be moved to various areas of the range are preferred with layer flocks. The shelter should provide about 1½ square feet of floor space per bird. It should be equipped with roosts, or slat floors with the slats located 1½ inches apart.

Normally, turkey poults can be put out on the range at 8 weeks of age. The flock should be well-feathered, especially over the hips and back, before they are put on range. Check the weather forecast for several days before putting the birds on range. It is best to move them in the morning. This helps to get the birds used to range conditions without losses.

A range area that has been free of turkeys for at least one year and preferably for two years, is desirable. Poorly drained soil on the range should be avoided. Stagnant surface water may be a factor in turkey disease outbreaks. A temporary fence can be used to confine the flock to a small area of the range. The fence can be moved once a week, or as often as the range and weather conditions indicate. Artificial shade should be provided if there is no natural shade. Several rows of corn planted along the sunny side of the range area provides good shade and will also provide some feed as it matures. About one acre of good pasture is required for 250 turkeys. If range shelters are used, it is well to move them every 7 to 14 days depending upon the weather and the quality of range. The feed and watering equipment should also be moved as needed to avoid muddy and bare spots.

It is possible to use a combination of confinement and range-rearing by providing a permanent shelter such as a barn with fenced-in range areas around the barn. These fenced-in areas can be used on an alternate basis every year or two.

Range crops will depend upon the climate, soil and the range management. Many turkey ranges are permanently seeded, others are a part of a crop rotation plan. As part of a 3 or 4 year crop rotation, legume or grass pasture and an annual range crop such as soybeans, rape, kale, sunflowers, reed canary grass and sudan grass, have been used successfully. Sunflowers, reed canary grass and sudan grass provide green feed and also shade. For permanent range, alfalfa, ladino clover, blue grass, brome grass and others are very satisfactory.

Where feeders are to be used outside, they should be waterproof and windproof so that the feed is not spoiled or blown away (See *Figure 21*). The feeders should be placed on skids, or be small enough so that they can be moved by hand. Trough-type feeders are inexpensive and relatively easy to construct. Specialized turkey feeding equipment can also be purchased. To prevent excessive feed waste, all feeding equipment should be designed so that it can be adjusted as the birds grow. The lip of the feed hopper should be approximately on line with the bird's back to prevent waste, and the feed level in the hopper should be kept at about half to prevent waste. Pelleted feeds are less likely to be wasted on range. Provide two 10-foot feeder troughs per 100 birds if the flock is hand fed each day. When bulk feeders are used, feeder space should conform to the equipment manufacturers recommendations. Provide one 4-foot automatic trough

(Figure 21. A Range Feeder.)

waterer or the equivalent per 100 birds. The waterers should be cleaned daily and disinfected weekly. If possible, the waterers should be shaded with portable or natural shade.

The range shelters should provide roosting space and cover for the birds during adverse weather and hot sun *(Figure 22)*. During the hot period of the year, particularly when birds are close to marketing time, it is advisable to pull extra shelters onto the range to provide more shade for the birds when natural shade is not available.

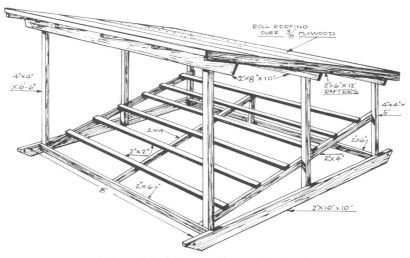

(Figure 22. A Turkey Range Shelter.)

debeaking

It is recommended that the poults be debeaked when they are from 2-5 weeks of age. Delaying beyond this period makes it difficult to handle the heavier birds and excessive feather picking may occur. To prevent cannibalism and feather pulling, debeaking should be regular practice before noticeable picking occurs.

The poults should not be debeaked at day of age at the hatchery, as this can interfere with the birds ability to eat and drink and may cause starveouts and dehydrated birds. The debeaking job is done with an electric debeaker and about ½ of the upper beak should be removed, being careful not to cut into the nostril. When birds are debeaked it is necessary to make sure that the feed and water levels are deep enough for the debeaked birds. See *Figure 23* for correct debeaked appearance of turkeys.

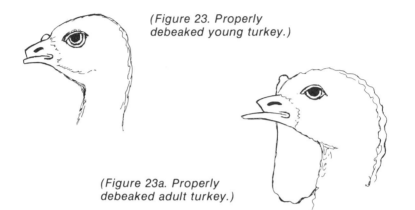

(Figure 23. Properly debeaked young turkey.)

(Figure 23a. Properly debeaked adult turkey.)

desnooding

The snood (a fleshy appendage on the top of the head back of the beak) is sometimes removed. Removal of the snood helps to prevent head injuries as a result of fighting or picking, and may prevent erysipelas infection from getting started in the flock. The snood may be removed at day old by pinching between the finger and thumb nail. It may be cut with finger nail clippers or scissors up to approximately 3 weeks of age.

wing clipping or notching

The wing feathers of one wing are sometimes cut off with a sharp knife to prevent the birds from flying, if on range. Wing notching, or the removal of the end segment of one wing with a debeaker, is another method sometimes used to prevent flying on range. Wing notching can be done with an electric debeaker from day of age up to 10 days of age. Wing clipping and wing notching are not used as much as they once were because of injuries resulting when the birds attempt to fly. This may cause carcass bruises and detract from the dressed appearance.

toe clipping

Toe clipping is frequently done to prevent scratches and tears of the skin on the bird's backs and hips. It is especially helpful where birds are crowded or nervous, but appears to help even when birds are on the range. The two inside toes are clipped so that the nails are completely removed. Surgical scissors or an electric debeaker may be used. This should be done at the hatchery.

general management recommendations

Young poults should be isolated from older turkeys. Care should be taken not to track disease organisms from the older stock to the younger stock. A good control program for mice and rats should be followed. These rodents are not only disease carriers but also consume large quantities of feed. Rats can kill young poults. Ideally, no other avian species such as chickens, game birds or waterfowl should be on the same farm. However, with the medications available today, it is possible to grow small flocks of turkeys where other types of birds are kept, if they are kept separate. It is still somewhat risky. If abnormal losses or disease symptoms occur, birds should be taken promptly to a disease diagnostic laboratory for diagnosis. One of the first symptoms of a disease problem is a reduction in feed and water consumption. Without good records on daily feed consumption it is difficult to detect changes in feed intake. Dead birds should be disposed of immediately in a sanitary manner, by incinerating or by use of a disposal pit.

With good management, the turkey grower should be able to raise 95% of the turkeys started to maturity. With high poult costs and feed costs, mortality can become very costly, especially when birds are lost towards the latter part of the growing period.

Depending upon the disease exposure in the given area it may be necessary to vaccinate the poults for such diseases as Newcastle, fowlpox, erysipelas, or fowl cholera. In planning a vaccination program for your flock, check with the poultry diagnosticians at your state laboratory, your County Extension Agent, or some other knowledgeable person. Frequently small flocks are not vaccinated.

Medication is effective in reducing losses from such diseases as coccidiosis and blackhead when used in the feed at preventive levels. Antibiotics and other drugs are of value in preventing and treating diseases, but such medications should not be used as a substitute for good management.

the production of turkey hatching eggs

Occasionally there are individuals who want to mate a few turkeys for the production of hatching eggs. The turkey breeding flock is not only somewhat more difficult to manage than the chicken breeding flock but is considerably more expensive. The cost of growing the bird to maturity is several dollars *(Table 18),* and the amount of feed required to maintain the hens and toms during the holding and breeding periods is also substantial *(Table 21)*.

TABLE 21

FEED CONSUMPTION OF TURKEY BREEDERS
(POUNDS PER BIRD PER DAY)

Type Turkey	Hens	Toms
Large	.60	1.00
Medium	.45	.75
Small	.35	.65

Source: *Turkey Production AGR Handbook No. 393 U.S.D.A.*

selecting the breeders

Those turkeys to be kept as breeders should be started about 8 months before egg production is desired. Those that come into production at too early an age lay small eggs. Fertility tends to be poorer in small eggs.

Select only the best birds as breeders. Look for candidates with good, full breasts, (those with no protruding keel bones). Select for strong, straight legs, straight keel bones, and straight backs. Use only those birds that are healthy and vigorous for breeders.

mating

One tom per 20 hens is adequate for the light type breeders. Medium turkeys can be mated at the ratio of 1 tom per 18 hens. Large turkeys require 1 tom for 16 hens. A few extra toms should be kept to replace poor mating birds or those that die.

lighting

The usual procedure is to commence lights on the toms about 4 to 5 weeks prior to mating. The toms need this light period to stimulate them to produce sperm. The hens should receive light about 3 weeks prior to the onset of egg production.

A 13 to 14 hour light period, artificial and natural combined, should be used. A 50 watt bulb for each 100 square feet of floor space is adequate. A time clock can be used to bracket the natural daylight hours with artificial light.

egg production

Turkeys are not as good layers as most chickens. Heavy strains may be expected to produce 40 to 50 eggs during the season. Medium turkeys lay 50-70 eggs per hen and the small strains 85 to 100 eggs per bird.

First year egg production is best. After the first year, production diminishes at the rate of approximately 20% each succeeding year. Egg size increases with age and hatchability tends to decrease.

mating habits

Females will not mate until they commence egg production. Toms must be ready for semen production when the hen is ready to mate. The toms strut most during the mating season. When the hen is ready to mate she approaches the tom of her choice. The hen squats near the tom and the male mounts her and copulation usually takes place. Some toms may never mount the hen, others may mount but not complete the insemination. When the male performance is not satisfactory, the hen may become disinterested and not mate for a period of time. If this happens, low-fertility can result.

fertilization

Fertilization of the egg takes place in the upper part of the oviduct. Sperm are stored there for several days to several weeks and can fertilize eggs up to 3 weeks after insemination. Storage life of the sperm shortens as the birds become older.

artificial insemination

Some breeder hens are fertilized totally by artificial insemination. Others are artificially inseminated on a supplemental basis to improve fertility where natural mating proves to be less than satisfactory.

The semen used for artificial insemination is obtained by milking the toms. Semen can be collected from the toms 2 or 3 times per week. The tom is stimulated by stroking its abdomen and pushing the tail upwards and toward the head. The male copulatory organ enlarges and partially protrudes from the vent. By gripping the rear of the copulatory organ with the thumb and forefinger from above and fully exposing the organ, the semen is then squeezed out with a short sliding downward movement. The males soon become trained and ejaculate easily when stimulated. The semen is collected in a small glass beaker or a stoppered funnel. The tom yields 2/10 to 5/10 of a cubic centimeter. The semen should be clear and free of fecal material. Some producers withhold feed and water from the toms for 8-12 hours to help avoid fecal contamination of the semen.

Semen should be used within 30 minutes. About 1/40 cc of semen per hen is adequate. Usually good fertility will result when hens are inseminated 2 times at 4 day intervals at the onset of egg production. Insemination every 2 or 3 weeks thereafter is usually desirable unless fertility is high in the mated flock.

The hen is inseminated by exposing the opening of the oviduct and inserting a small syringe without needle into the oviduct about 1½ inches. The oviduct is exposed by pushing outward and exerting pressure on the abodmen, while at the same time forcing the tail upward toward the head. The oviduct can be exposed or protruded only in those hens that are in laying condition.

broodiness

Broodiness is an inherited characteristic. Turkeys are inclined to be more broody than most breeds of chickens. Some strains of turkeys are more broody than others. It is difficult to identify broody turkeys. If the birds are found on the nest in the early morning or evening it is safe to assume they are broody. Broody birds should be driven into a broody pen or yard and left there for about five days. The birds should be fed and watered in the broody pen. Dark spots and areas where they can rest should be eliminated. No nests should be provided. Slat or wire floors are best for broody pens. One or more toms in the pen will tend to keep the hens on their feet and discourage broodiness. Hens usually want to mate a few days after they go broody so this practice may also improve fertility.

Broodiness may be discouraged by moving the birds to different areas, by providing roosts and chasing birds off the nest when the eggs are gathered. Gathering eggs several times each day may also help prevent broodiness.

housing

Buildings which are used for brooding young stock are satisfactory for breeders. Confinement is preferred to range management where unfavorable weather is common or where there are predators that may be a problem. The houses should be well insulated and ventilated for maximum comfort in cold or warm weather. The floor can be cement, asphalt or wood. It should be covered with a good litter. Houses should be thoroughly cleaned and disinfected between flocks.

More floor space is required for breeders. Usually 6 to 8 square feet per bird is recommended. Where only females are housed 5 to 6 square feet is adequate for large turkeys and 4 to 5 for small ones.

In warm climates or where winters are mild, breeders may be given access to fenced range or yard. They should have a shelter to protect them from bad weather and predators. The range area should be well drained and provide about 150 square feet per bird. Where a yard is used, allow 4 to 5 square feet of area per bird. The shelter should be equipped with roosts and provide at least 4 square feet of floor space per bird. The feeders, waterers and the broody pen are best located inside the shelter.

nests

Provide one nest for every 5 birds. Nests should be located in a dark area of the house. They should be available prior to the time egg production starts so the hens will get used to them.

Some breeders prefer tie-up or trap nests but open type nests may be used. Nests should be 2′ x 2′ if they are the manger or open type. If nests are placed outside the shelter they should be protected by a roof.

feeding

Breeders should be put on a special turkey breeder diet one month before egg production is expected. Feed according to the manufacturers instructions. All-mash diets are preferred but mash and grain diets are satisfactory if fed in the right proportions. If too much grain is fed in relation to the mash, egg production and hatchability may be reduced. If used properly this method has the advantage of making use of home-grown grain without having to grind or mix it. Home-grown grains can be used in an all-mash program by grinding and mixing with a concentrate at home or having them mixed elsewhere.

Mash or concentrate may be in a pelleted form. There is less waste with pelleted feeds, particularly if fed on range. An insoluble grit or gravel should be available for confinement breeders. When the birds are fed and watered out-of-doors the equipment should be moved frequently to avoid muddy spots and possible sources of disease.

Six linear inches of feeder space should be provided for large turkeys and 4½ linear inches for the small birds.

Provide two automatic waterers per 100 birds. One linear inch of water space per hen is adequate. Turkeys normally consume 2 pounds of water for every pound of feed. Lack of water can be more harmful to the birds than the lack of feed.

care of eggs

Eggs should be gathered at least 3 times per day, more frequently if the birds tend to crowd certain nests. Frequent gathering will help to avoid breakage, dirty eggs and possibly frozen eggs.

If eggs are cleaned they must be cleaned properly, using a detergent sanitizer made for egg washing. A washing temperature of 110° to 115° F. is recommended. Excessively dirty eggs gathered in hot, wet weather may easily become contaminated. Contaminated eggs often explode in the incubator.

If eggs are to be held before incubating they should be turned daily. When eggs are turned and held under good storage conditions hatchability may be good for two weeks. Eggs can be placed in filler-flats or cartons. The container can be elevated at one end to slant the eggs about 30°. At least once each day, shift ends with the container and elevate the opposite end. Hatchability will be improved if this is done.

flock health

The hens should be debeaked when placed in the breeding pens. Remove about one-half of the upper beak *(Figure 23)*. Feed and water levels should be maintained at a height that will permit the debeaked birds to eat and drink.

Some states require that breeders be blood tested for Salmonella pullorum and Salmonella typhoid and possibly other diseases such as infectious sinusitis, paratyphoid and PPLO (pleuro-pneumonia-like infection.) Follow the recommendations of your County Agent, State Extension poultry Specialist or diagnostic laboratory.

Bronze Turkeys

Broad Breasted White Turkeys

Source: *Poultry Tribune, Mount Morris, Illinois.*

10

WATERFOWL PRODUCTION

Waterfowl can be raised with relative ease. They are hardy and not subject to as many diseases as are other types of poultry. Numerous small flocks of waterfowl are grown to grace a small pond, as a hobby, or for exhibition. Not to be overlooked is the fact that properly finished and dressed, waterfowl are excellent for food. A number of people, in fact, prefer waterfowl to other types of poultry for the Thanksgiving and Christmas holidays.

Breeds Of Waterfowl

ducks

There are several breeds of ducks available. The breed one selects will depend upon the purpose for which they are to be raised. If ducks are to be kept for meat purposes, then the best choices include the White Pekin, the Aylesbury, the Rouen and the Muscovy ducks. If they are to be kept for ornamental purposes, or for hobby or exhibition purposes, the small ducks may better fit the bill. Included in this group are the White and Grey Calls, the Wood duck, Mandarin, and Mallard. If the interest is in a breed of ducks which produce eggs, then the choice should be the Khaki Campbells or the Indian Runners.

Meat Breeds

white pekin

The White Pekin is well suited for the production of meat. It produces good quality meat and reaches a market weight of about 7 pounds in approximately 8 weeks. It is a large, white feathered bird. Its bill is an orange-yellow color, its legs and feet a reddish-yellow color and it has a yellow skin. The adult drake weighs 9 pounds, the adult duck 8 pounds.

The White Pekin is a fairly good egg producer. The average yearly egg production reaches approximately 160 eggs. It is a poor setter and seldom raises a brood of ducklings.

aylesbury

The Aylesbury originated in England and, like the White Pekin duck, is a good meat bird. It reaches market weight in about 8 weeks. The Ayelsbury has white feathers, white skin, a flesh colored bill, light orange legs and feet. The eggs are tinted white. The adult drake weighs 9 pounds, the adult duck 8 pounds. It lays somewhat fewer eggs than the White Pekin and is also a poor setter.

muscovy

There are several varieties of Muscovies, the white being the most desirable for market purposes. They produce meat of excellent quality and taste, when marketed before 17 weeks of age. They are relatively poor egg producers but good setters. The adult drakes weigh 10 pounds, the adult ducks 7 pounds. Muscovies have white skin.

rouen

The Rouen duck reminds one of the Mallard. It has the same striking color patterns as the Mallard but is much larger. The adult drake weighs 10 pounds, the adult duck 9 pounds. Its pigmented plumage gives it a less desirable dressed appearance. It is excellent for home consumption, where the dressed appearance is not so important.

Egg Producing Breeds

khaki campbell

The Khaki Campbells were developed in England. Several varieties of the Khaki Campbell ducks have been selected for high egg production. The egg production of some of these varieties is reported to average close to 365 eggs per duck in a laying year. It is interesting to note that this is a higher rate of production than the highest producing strains of chickens.

The male has a brownish bronze lower back, tail converts, head and neck, and the rest of her plumage is khaki. The beak is greenish black and the legs and toes brown. The adults weigh only 4½ pounds, thus the Khaki Campbell is not known for the production of meat.

the indian runners

Although the Indian Runners originated in the East Indies, their egg production qualities were developed in Europe. They are second only to the Khaki Campbell in egg production. There are three Indian Runner varieties, the White Penciled, Fawn, and White. All of these varieties have orange to reddish-orange feet and shanks. The males and females weigh approximately 4½ pounds. They are not good meat birds.

Pekin Duck

Colored Muscovy Ducks

Source: *Poultry Tribune, Mount Morris, Illinois*

Breeds Of Geese

The most popular breeds for meat production are the Toulouse, Emden and African geese. Other common breeds in the United States include the Chinese, Canada, Buff, Pilgrim, Sebastipole and Egyptian.

In choosing a breed one should consider the purpose for which they are to be raised. Geese may be raised for either meat or egg production or as weeders for show birds, or as farm pets. Some of the cross breeds, such as the cross between the white Chinese male with the medium sized Emden female, usually results in fast growing white geese of good market size.

toulouse

The Toulouse has a broad, deep body and is loose feathered, a characteristic which helps give it its large appearance. The plumage is dark grey on the back, gradually shading to a light grey edge with white on the breast and white on the abdomen. The bill is pale orange and the shanks and toes are a deep reddish orange.

emden

The Emden is a pure white goose. It is a much tighter feathered bird than the Toulouse and, therefore, appears more erect. Egg production averages from 35-40 eggs per bird. The Emden is a better setter than the Toulouse and is one of the most popular breeds for meat. It grows rapidly and matures early.

african

The African has a distinctive knob on its head. The head is a light brown, the knob and bill are black and the eyes dark brown. The plumage is ash brown on the wings and back and a light ash brown on the neck, breast and underside of the body. It is a good layer, grows rapidly and matures early. It is not as popular for meat production because of its dark beak and pin feathers.

chinese

The Chinese are smaller than the other standard breeds. There are two varieties, the brown and the white. Both varieties mature early and are better layers than are the other breeds. They average from 40-65 eggs per bird annually. The Chinese grows rapidly. It is a very attractive breed and makes a desired medium sized meat bird. It is very popular as an exhibition and ornamental breed.

canada goose

The Canada Goose is the common wild goose of North America. The weight of the Canada Goose ranges from about 3 pounds for the lesser Canada Goose to about 12 pounds for the giant or greater Canada Goose. The Canada is a species different from other breeds of geese. It can be kept in captivity only by close confinement, by wing-clipping or pinioning the wings. They may be kept only by permit which must be obtained from the Fish and Wildlife Service of the United States Department of Interior in Washington, D.C.

The breed does not have the economic value of other breeds of geese, they mate only in pairs, are late maturing, and lay very few eggs.

pilgrim

The Pilgrim is a medium sized goose and is good for meat production. The males and females of this breed can be distinguished by feather color. In day old goslings the male is a creamy white and the female is grey. The adult male remains all white and has blue eyes. The adult female is grey and white and has dark hazel eyes.

buff

This has only fair economic qualities as a market goose and only a limited number are raised for meat. The color varies from the dark buff on the back to a light buff on the breast and from light buff to almost white on the under part of the body.

sebastipole

The Sebastipole is a white ornamental goose which is very attractive because of its soft plume-like feathering. The breed has long curved feathers on its back and sides and short curled feathers on the lower part of the body.

egyptian

This is a long-legged but very small goose. It is kept primarily for ornamental or exhibition purposes. Its coloring is mostly grey and black with touches of white, reddish brown and buff.

Getting Started

Those who wish to raise a small flock of ducks or geese can best get started by the purchase of day-old ducklings or goslings. This eliminates the need for keeping breeding birds and incubating the eggs. Day old stock is available in most areas of the country. A list of those hatcheries selling waterfowl is printed in the National Poultry Improvement Plan, Hatchery list. This can be obtained by writing to the United States Department of Agriculture in Washington, D.C. Other good sources of information concerning the availability of young birds include the state Extension Poultryman and the County Extension Agent in your area.

brooding and rearing

As is the case with other types of poultry, waterfowl need to be put on feed and water as soon as they are placed under the brooder. Young ducklings, and particularly young goslings are somewhat more hardy than are chickens and turkeys. However, they do need to be kept where it is warm, dry, and free of drafts. Most any type of building that will provide these conditions is suitable for brooding ducklings or goslings. The building should be designed so that it can be ventilated without chilling the birds.

The brooder house should be cleaned and disinfected well before the time the birds are to arrive. When they arrive, the house, the litter, the feeding and watering equipment and the brooder stove should be ready. The brooder should be operating at least one day before the birds arrive. The first day a brooder guard should be placed approximately 2 feet from the hover. It should be gradually moved back and entirely removed at about 7 days. The brooder guard will confine the birds to the brooding area and prevent huddling and chilling before they are able to adjust to the location of the heat source.

Waterfowl may be started on wire, slat or litter floors. The most practical method is to start them on litter floors. Wood shavings, sawdust, chopped straw or peat moss are all good litter materials. The material to be used should be free of mold, since moldy litter can cause mortality. The proper equipment arrangement for brooding is shown in *Figure 10*.

The birds should have adequate floor space. Ducklings need ½ square foot of floor space per bird the first week, ¾ of a square foot the second and 1 square foot the third week. If they are to be housed in confinement this space allotment will need to be increased to 2½ square feet by the time they reach 7 weeks of age. Goslings will need more floor space. They should have ½ to ¾ of a square foot of floor space the first week and 1 to 1½ square feet the second week. The amount should be increased slightly until they go on range. They can be placed on range when 2 to 4 weeks of age. No shelter is needed but some type of shade should be provided. Ducklings normally need supplementary heat for approximately 4 weeks. During warm weather or in warm climates, heat may be needed for just the first two or three weeks. Goslings usually can get by without heat after 2 weeks of age. Waterfowl feather very rapidly and thus do not require as long a brooding period as do baby chicks. Any brooder unit suitable for brooding chicks or turkeys is satisfactory for ducklings or goslings. Electric, gas, oil, coal or wood burning brooder units are all suitable.

Infrared heat lamps are excellent for brooding small groups of birds. One infrared lamp for up to 30 birds is satisfactory. To determine the number of ducklings or goslings a hover type brooder will accommodate, cut the brooder's rated chick capacity in half. They require 13 to 14 square inches of hover space per bird.

The brooder temperature should be in the vicinity of 85-90° F. the first week, and reduce approximately 5° per week during the next few weeks. The behavior of the young birds is the best guide to the temperature required (See *Figure 11*). When the temperature is too hot, the birds will crowd away from the heat. High temperatures may result in slower weight gains and slower feathering. When the temperature is uncomfortably cold the birds will tend to huddle together under the brooder or, perhaps, crowd into corners. When a hover-type brooder is used, a night light will tend to discourage crowding; where infrared brooders are used, enough light is provided to make extra light unnecessary. When the temperature is just right the birds will be well distributed over the floor and using all the feeders and waterers.

Keep clean drinking water in front of the birds at all times. Hand-filled water fountains or automatic waterers may be used. Place the water fountains on wire or slatted platforms or over a screened drain. This will help to keep the litter dry. The waterers should be of the type that the birds cannot get into. Waterers should be cleaned daily.

Waterfowl should have access to range or a yard when old enough to tolerate the weather conditions. Ducklings will manage nicely on range at 4 weeks of age unless the weather is cold. Goslings can be placed out-of-doors at 2 weeks of age weather permitting. They need shade and cannot tolerate chilling rains until they are well feathered on the back. When birds reach 5-8 weeks of age they need shelter only during extreme weather conditions. Most commercial growers provide ponds for their birds at approximately 5 weeks of age. However, waterfowl can be raised without swimming water.

As with other types of poultry, feather pulling can be a problem with ducks. If this vice starts, the birds should be given additional space. It may be necessary to debill the birds if the problem continues. This is done by nipping off the forward edge of the upper bill with an electric debeaking machine.

feeding and watering space

Several types of feeders are satisfactory for waterfowl. Feed hoppers, such as those used for chickens and shown in *Chapter 2, Figures 3* through *3e,* work well. Small feeders can be used until the birds are two weeks of age. Large feeders should be used for older birds and breeding stock. Feed hoppers that are used outdoors should be covered to protect the feed from wind and rain. A range feeder designed for chickens or turkeys will work well. A plan for a range feeder is also shown in *Chapter 2.*

To start ducklings and goslings off it is well to use chick box covers, shallow pans, or small chick feeders. Hanging tube-type feeders are excellent. One of these with a pan circumference of 50″ is suitable for 50 ducklings or 25 goslings for the first two weeks.

Pans or troughs are satisfactory waterers. They should be equipped with wire guards or grills to prevent the birds from playing in them and spilling water. The waterers should be placed over low wire covered frames, and preferably, over drains to prevent the litter from getting too wet. A water stand which is satisfactory for waterfowl is shown in *Figure 5*. Increase the size and number of waterers as the birds become older. Approximately 4' of trough is good for 250 young waterfowl for the first few weeks. The waterer should be wide enough to permit the bird to dip its bill and head into it.

feeding

Some commercial feed companies formulate rations specifically for ducks and geese. However, in a number of areas specially prepared rations will not be available. It is possible to mix your own feed. But before doing so, write to your Extension Poultry Specialist for formulas. If feeds for ducks are not available, they may be started on crumbled or pelleted chick starter for the first 2 weeks. Pellets are recommended because they are easier to consume. They also reduce waste, and do not blow around like mash when used out of doors. Feed conversion is usually better with pellets. However, the lack of pelleted feed should not discourage a grower who wishes to produce ducks or geese on a small scale. Satisfactory results can be obtained with mash. If a chick starter is used, it is desirable that the starter not contain any drugs which may be harmful to the ducklings. The use of a medicated feed is not usually necessary. Coccidiosis is not nearly the problem in ducklings that it is in chickens, though it does affect some flocks.

Ducklings and goslings do well on a starter diet that contains 22% protein. At 2 weeks of age thay can be changed to a 15 to 18% protein diet. If a special grower diet is not available a chick grower is satisfactory. When ducks are to be dressed as green ducks, that is at 7-9 weeks of age, this diet can be fed for the full period. In addition to a pelleted grower ration, cracked corn or other grains are often included in the diet for goslings. Feed should be kept before the birds at all times, and an insoluble grit provided. Grit enables the gizzard to grind fibrous material so that it can be passed on and further digested in the intestine. Ducks are not as good foragers as geese but may be put on pasture at 4 weeks of age if to be kept beyond the 7 to 9 week period. Feed consumption and liveweight information for ducks and geese are shown in *Tables 22* and *Table 23*.

TABLE 22

AVERAGE LIVEWEIGHT, FEED CONSUMPTION AND FEED CONVERSION RATIOS OF WHITE PEKIN DUCKLINGS AT DIFFERENT AGES MIXED SEXES

Age	Liveweight	Feed Consumption		Feed/lb of weight gain to date
		Weekly	Cumulative	
weeks	lb.	lb.	lb.	lb.
1	0.60	0.50	0.50	0.83
2	1.68	1.64	2.14	1.27
3	2.98	2.55	4.69	1.57
4	4.01	2.55	7.24	1.81
5	5.13	3.27	10.51	2.05
6	6.19	3.57	14.08	2.27
7	6.96	3.87	17.95	2.58
8	7.54	3.39	21.34	2.83

Source: *Cornell University*

TABLE 23

GROWTH RATE AND FEED CONSUMPTION OF WHITE CHINESE, EMBDEN GOSLINGS

	Confinement-reared			Range-reared		
Age	Avg wt per gosling	Cumulative feed consumption per gosling	Feed per lb. of gosling to date	Avg wt per gosling	Cumulative feed consumption per gosling	Feed per lb. of gosling to date
weeks	lb	lb	lb	lb	lb	lb
3	3.3	5.25	1.75	3.3	5.25	1.75
6	8.1	17.32	2.22	7.8	12.60	1.68
9	10.8	34.92	3.32	10.1	19.97	2.03
12	12.3	47.54	3.96	11.6	30.62	2.71
14	12.8	56.09	4.48	11.8	38.10	3.31

Source: *Duck and Goose Raising, Ministry of Agriculture and Food — Ontario*

The Breeding Flock

If a breeding flock is to be kept the prospective breeders may be selected at about 6-7 weeks of age. A few extra should be selected at that time to allow for culling in the future. Birds selected for breeders should then be placed on a breeder developer diet, a diet containing less energy than the starter or grower diets. Frequently the breeder developer diet is fed on a restricted basis to prevent the birds from becoming too fat. Birds that are too fat will lay fewer and smaller eggs. Where a diet is fed on a restricted basis, have plenty of feeder space or spread the pellets over a wide area to make sure that all of the birds get their share.

To obtain optimum fertility, it is best to feed a special breeder diet. Breeder birds should be switched to the special diet a month prior to the date of anticipated egg production. If a breeder developer diet is not available, a growing ration can be fed during this time.

selecting and mating the breeder flock

Ducks to be used as breeders are usually selected from the spring hatched birds. They may be selected at 6 to 7 weeks of age. It is possible at that time to differentiate the males from the females by their voices. The females honk and the males belch. (See the section on sex differentiation for further information). The ratio of drakes to ducks is normally 1 to 6. A few extra should be selected to take care of mortality or to permit further selection during the growing period.

Breeders should be selected for vigor, body weight, conformation, and feathering as well as breed characteristics. Although ducks demonstrate some selectivity in mating they are essentially polygamous in their mating habits.

Geese are somewhat different in their mating habits. Some breeds tend to be monogamous, the Canada Goose being an example. Layer breeds mate better in pairs or trios, and the ganders of some of the smaller breeds may accept 4 or 5 females. When selecting geese for the breeding flocks, select for body size, rate of growth, liveability, egg production, fertility and hatchability as well as breed characteristics.

housing and managing the breeder flock

Housing for waterfowl does not have to be as tight as that for chickens and turkeys. The pen should be well lighted, well ventilated and supplied with plenty of dry litter. Waterfowl, and especially geese, prefer to be outside even during severe winter weather. Insulated buildings are not too important. A simple shed, a small house or an area within an existing barn can be used as a breeder pen. If more than one trio occupies a shelter it should be divided to prevent fighting.

Dirt floors may be used but cement or wood floors are easier to clean and disinfect. Litter materials that are used for brooding are satisfactory for the breeders. The litter should be kept dry. Wet and caked areas should be removed and dry litter added from time to time as required. Floor space requirements vary depending upon the type of bird and whether or not a yard is available. Ducks in confinement need 5 to 6 square feet per bird and 3 square feet when a yard is provided. Geese having access to a yard should have 5 square feet per bird in the house. Up to 40 square feet per bird should be provided in the yard.

Windows and doors should be left open during the daylight period to allow adequate circulation of air. Good ventilation is also needed at night to prevent over-heating during hot periods and to reduce condensation during the winter. Supplementary heat is not necessary.

The birds will make their own nests in the litter and lay about anywhere in the house. They prefer a somewhat secluded spot. Simple nest boxes can be provided in rows along the wall. The nests should be 12″ wide, *18″* deep and 12″ high. Nests for large geese should be 24 inches square. Some prefer to not partition the nests at all. The tops and fronts of the nests are open. Straw and wood shavings make good nesting materials. The nesting material should be kept clean to prevent dirty and soiled eggs. One nest is adequate for every 3 to 5 breeders.

The pens may be kept drier if the watering devices are placed outside the house. Ducks and geese can go without water overnight, provided they do not have access to feed. When permitted to eat without available water they can choke to death on the feed. If the birds are locked in the house overnight with no water, the feeders should be empty or closed. If the birds are watered in the house, again, the water supply should be placed above a screened drain.

Yards should slope gently away from the house to provide good drainage. Mud holes and stagnant puddles of water are sources of diseases and parasites. Manure will accumulate in the yard, so it should be cleaned occasionally. The frequency of cleaning will depend upon the number of birds and the size of the yard. Large flocks are normally provided 75 square feet of yard space per bird.

Ducks have the tendency to run in circles if startled in the dark. This is a more serious problem with large flocks. An all night light should be used in the house if stampeding is a problem. One 15 watt bulb for each 200 square feet of floor space is adequate to prevent this.

The breeders should not be brought into full production before they reach 7 months of age. If production starts sooner there will be a problem of small eggs and low hatchability. Fourteen hours of light will stimulate them to produce in the fall and winter months when the days are short.

Females should be lighted three weeks before production. The males should be prelighted 4 to 5 weeks prior to mating. Pre-lighting will help fertility. The lamps used for preventing stampeding in duck pens will also stimulate egg production. However, larger bulbs give better light stimulation.

Ducks are much better layers than geese. They lay more eggs over a longer period of time. Geese tend to lay every other day but may lay 2 or more successive days. Daily gathering of eggs will help to discourage broodiness (the urge to set) and an accompanying pause in production. Separating broody birds from their mates and confining them with feed and water will sometimes discourage broodiness.

The birds should reach a peak of production at approximately 5 or 6 weeks after the onset of lay. The rate of production will depend upon the breed, the feeding program and the management they receive. Older geese make better breeders than the young ones. The geese lay more eggs their second year and 2 year old ganders tend to yield better fertility. Geese will lay until they are about 10 years old. Ganders may be kept as breeders for 5 years.

egg gathering and care

Ducks lay most of their eggs during the night and early morning hours. The eggs should be collected early so as to prevent them from getting soiled and cracked. The birds should be kept in the house each day until the eggs are laid. Eggs that are laid outside may freeze during severely cold weather, killing the embryos. The eggs can also be badly soiled if laid outside. Eggs that are dirty should be washed soon after gathering. The wash water should be warmer than the eggs. The recommended wash water temperature is 115° F. If cold water is used it will cause the egg contents to contract, and bacteria and other organisms may be drawn through the pores of the shell and contaminate the egg contents. This can cause the eggs to explode in the incubator. The eggs should be washed in a detergent sanitizer which may be obtained from an agricultural supply house. The washing time is 3 minutes. The maufacturer's directions for use should be followed.

Badly misshaped, abnormally small or large eggs, or eggs that have been cracked, should not be saved for incubation. Chances are they will not hatch. The eggs should be stored at a temperature of 55° F. and a relative humidity of 75%. If hatching eggs are to be stored for more than a week prior to incubation, they should be turned daily. This will prevent the yolks from sticking to the shell membranes causing a reduction in hatchability. Eggs should be stored on egg filler flats or egg cartons. One end of the container may be propped up at an angle of 30°. Each day change ends with the container. Eggs that are stored for 2 weeks or longer decline in hatchability quite rapidly.

incubation

The incubation period for most domestic ducks is 28 days. The Muscovy requires 35 days. The Canada and Egyptian geese require 35 days, all other geese 30 days.

It is possible to incubate waterfowl eggs by natural means in small flocks. A goose will cover 9 or 10 eggs and a duck 10 to 13 eggs. While the Muscovy duck is a good setter, most other breeds of ducks are not. If available, a broody hen may be used and will cover 4-6 goose eggs, or 9-11 duck eggs. The nest should be placed where the hen won't be disturbed during the incubation period. A nearby source of feed and water should be provided for the setting hen. Where hens are used to incubate waterfowl eggs, the early hatched should be removed from the nest as soon as they are hatched and placed under a brooder. This will prevent the hen from leaving the nest before completion of the hatch, or trampling some of the young if she becomes restless.

The nest may be placed on the ground, or inside a building. Turkey hens and Muscovy ducks are larger and better than chicken hens for hatching duck and goose eggs. Ten to twelve goose eggs can be set under a turkey or duck, depending upon the size of the bird. Hens should be treated for lice before the eggs are set.

When chicken or turkey hens are used for setting, it is necessary to add moisture. The eggs should be sprinkled with luke warm water during the incubation period and the nest and straw should be placed on the ground or on a grass covered turf. This helps to increase the moisture. Some growers indicate that if the eggs are lightly sprinkled, or dipped in lukewarm water for half a minute daily during the last half of the incubation period they hatch much better. If ducks or geese have access to water for swimming no additional moisture is needed. If a setting hen does not turn the eggs they should be marked with a crayon or pencil and turned daily by hand. Usually the hen will turn them.

Artificial incubation is used extensively for hatching waterfowl eggs. In the large forced-draft machines a temperature of 99¼ to 99¾° F. is usually used. For the incubation of duck eggs the wet bulb thermometer reading during the incubation period should be 86-88° F. and during the hatching period 92° F. Since the humidity requirements are so much higher for waterfowl eggs than for chicken eggs, the two should not be set in the incubator at the same time.

In the small still air incubators the temperature should be 100½° F, 101½°F, 102½°F and 103°F respectively, for the first, second, third and fourth weeks of incubation. However, satisfactory results may be obtained by running the machine at an even 102° for the entire period. The thermometer reading should be taken at the top of the egg. The eggs in a still-air machine are set in a horizontal position, usually, and are turned 180 degrees on each turn. The eggs should be turned 3-4 times daily. The moisture requirements are the same for goose and duck eggs.

The eggs may be candled 4 to 6 days after incubation starts. Candling is done by passing the egg over an electric candling light in a dark room. The living embryo of the egg will appear as a dark spot at the large end of the egg near the air cell. From this dark spot radiates blood vessels giving a spider-like appearance. Embryos that have died will appear as a spot stuck to the shell membrane and no radiating blood vessels will be apparent. Infertile eggs will appear clear. The infertile egg should be removed from the incubator, along with any cracked eggs that are detected.

Usually the eggs are candled three days prior to hatching. At this time, the fertile eggs will appear dark, except for the large air cell end. The infertiles will be clear. Turning of the eggs should be discontinued three days prior to the end of the incubation period. From that time to hatching, the eggs should not be disturbed other than to open the incubator to add water.

Since the hatching period for waterfowl tends to be spread out over several hours the large commercial hatcheries frequently take the birds out of the machines as they are hatched and dried. However, it is necessary to prevent chilling of the newly-hatched watefowl and it may be better for the small producer to keep the incubator closed until the hatch is complete.

Some ducklings appear to need help from the shell. Conditions in small machines are sometimes difficult to control so this practice may be justified. However, birds that are helped from the shell probably should not be used for breeding stock as heredity may be a factor responsible for this condition.

sex differentiation of waterfowl

In general, the determination of sex in ducks is somewhat easier than for geese. The drakes usually have larger bodies and heads and the males have a soft throaty quack, while females have a loud, distinct, harsh quack. Also the main tail feathers on the mature drake are curled forward and not on the ducks. In colored breeds of ducks, the males are more brilliantly colored than are the females.

Characteristics which are used to distinguish between the adult male goose and the female goose are the longer neck and a somewhat coarser and larger head, as well as the larger body of the male. There are differences in the voice between sexes, though these are somewhat difficult to determine. The African and Chinese geese have knobs on their heads. The knobs of the male are larger than those of the female. In immature stock these differences are hard to distinguish. Even in the mature geese, there are cases where there are oversized females and undersized males. The Pilgrim breed can be sexed from day-old to maturity by the use of plumage color. The color of the male is light or white and the female is a dove grey.

sexing goslings and ducklings

There are occasions when it becomes desirable to be able to sex goslings and ducklings at an early age. The vent sexing of day-old ducklings and goslings is very similar. The sexing is done in a warm room using a strong light, so that the reproductive organs can be easily seen. (See *Figures 24, 25* and *26*).

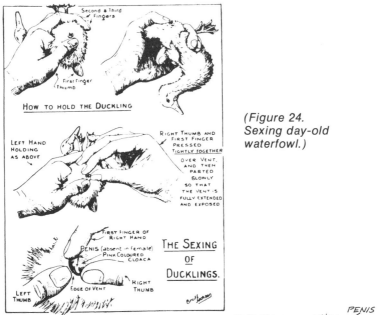

HOW TO HOLD THE DUCKLING

Second & Third Fingers

First Finger Thumb

LEFT HAND HOLDING AS ABOVE

RIGHT THUMB AND FIRST FINGER PRESSED TIGHTLY TOGETHER OVER VENT, AND THEN PARTED SLOWLY SO THAT THE VENT IS FULLY EXTENDED AND EXPOSED

FIRST FINGER OF RIGHT HAND

PENIS (absent in female)

PINK COLOURED CLOACA

THE SEXING OF DUCKLINGS.

LEFT THUMB

EDGE OF VENT

RIGHT THUMB

(Figure 24. Sexing day-old waterfowl.)

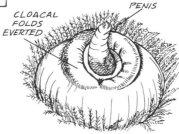

CLOACAL FOLDS EVERTED

PENIS

(Figure 25. Appearance of the male genital organ.)

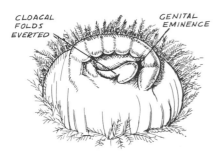

CLOACAL FOLDS EVERTED

GENITAL EMINENCE

(Figure 26. Appearance of the female genital organ.)

sexing adult geese

The bird is held over the bended knee or a table — on its back — with the tail pointed away from you. The tail end of the bird is moved out over the edge of the knee or the table so it can be readily bent downwards. Then the index finger is inserted into the cloaca about a half an inch and moved around in a circular manner several times to enlarge and relax the sphincter muscle which closes the opening. Then pressure is applied directly below and on the sides of the vent to expose the sex organs.

range for geese

During warm weather, geese can be put on range as early as the second week. A good share of their feed can be forage after they are 5-6 weeks of age. Geese are better foragers than are ducks. They tend to pick out the young tender grasses and clovers. They reject alfalfa and narrow leafed tough grasses and select the more succulent clovers and grasses. Geese should not be fed wilted poor quality forage.

It takes approximately an acre of range to support 20-40 geese, depending upon their size and the quality of the pasture. A 3 foot woven-wire fence will confine the geese to the grazing area.

Where good quality pasture is available, the amount of pelleted ration may be restricted to about 1 to 2 pounds per goose, per week, until the birds are 12 weeks of age. The amount of feed should be increased as the supply of good forage decreases, or when the geese reduce their consumption of grass. If the birds are being grown to market as meat, they should receive pellets on a free-choice basis after they are 12 weeks of age, even though on range. Geese may be fed pellets, mash or whole grains. For the first 3 weeks they should receive a 20-22% protein goose starter preferably in the form of pellets. Less waste occurs with pelleted feeds. After 3 weeks of age feed a 15% goose grower in the form of pellets. If a specialized grower diet or starter diet for geese is not available then the equivalent chick starter and grower can be utilized. Again, the diets need not contain coccidiostats or other medications.

Mash or whole grains can be fed alone, or they can be mixed at a 50-50 mash to grain ratio. At 3 weeks of age a mash to grain ratio of approximately 60-40 may be used. The proportions should be changed gradually during the growing period until, at market age, the geese are receiving a 40-60 ratio of mash to grain. Depending upon the quality and quantity of available range, these ratios may be adjusted up or down slightly. For maximum growth it is important that mash and grain mixtures provide a nutrient intake of approximately 15% protein, the same as the all-mash diet. Insoluble grit should be available to the geese throughout the growing period.

geese as weeders

Geese will eat many noxious weeds and are thus used as weeders for certain crops without seriously harming the crop. Geese are good weeders in orchards and for crops such as strawberries, sugar beets, corn, cotton and nursery stock.

For best results the goslings should not be more than 6 weeks old when started as weeders. Shade should be provided and waterers spaced throughout the field. The weeder goslings should be kept hungry. A light feeding of grain at night is enough and the amount is varied, depending upon the availability of weeds and grass in the crop which is being weeded.

If placed in a field before the weeds get a head start, 6-8 goslings properly managed can control the weed growth on one or possibly two acres of strawberries. After the fruit begins to ripen it is best to move the goslings to other crops, or put them on range, and fatten for market. A fence 30-36″ high serves to confine the goslings to the area to be weeded. Any pesticide, or insecticides that are used on adjoining tree crops, should not contaminate the weeds and grasses that are to be eaten by the goslings.

feathers

Duck and goose feathers have value. If properly cared for they may be a source of extra income, or they may be used at home. The feathers are used mainly by the bedding and clothing industries. It usually requires 5 ducklings or 3 goslings to produce one pound of dry feathers.

The feathers may be sold to specialized feather processing plants or a small producer can wash and dry them for home use. The feathers may be washed by using a soft lukewarm water to which has been added either a detergent or a little borax and washing soda. Following the washing, rinse, wring the moisture out and spread out to dry.

11

HOME PROCESSING
OF EGGS AND POULTRY

Egg Processing

E ggs, as they come from the nest, vary in several respects. They vary in size, shape, cleanliness, shell texture, interior quality, and some will be cracked and checked. If the right management, the right environmental conditions, the proper feed has been provided a healthy flock, then the majority of the eggs will be of good quality. If the family consumption of eggs keeps up with the production of the birds, then there won't be too many problems with quality differentiation or quality preservation. However, on occasion some flocks produce more eggs than can be consumed by the family and some of them have to be sold. Eggs are sold on the basis of size and quality, and thus it is necessary to understand a little bit about egg quality determination and egg size requirements.

egg quality

The quality of an egg never improves after it is laid. Quality diminishes with time and the rate depends upon how the eggs are treated. To best understand what happens to egg quality, a knowledge of the structure and the various parts of an egg is essential.

There are four basic parts to an egg, namely the shell, the shell membranes, the albumen and the yolk *(Figure 28)*. Fastened to the shell are two membranes. It is between these two membranes that the air cell is formed usually at the large end of the egg. When the egg is first laid this air cell is non-existent, but with the cooling and contraction of the egg contents, the air cell begins to appear a short time after the egg is laid. Since the shell is porous, moisture can escape from the egg, so the air cell enlarges as the egg is held for a period of time. The rate at which the air cell increases in size depends upon the holding conditions.

A relative humidity of 75% in the cooler will keep the evaporation to a minimum. The yolk of a high quality egg is very near the center of the egg. On the surface of the yolk is a small white spot known as the germinal disc or germ spot. It is at this point that fertilization takes place. Whether the egg is fertile or infertile, the germ spot looks exactly the same. In other words, it is not possible to tell with the naked eye whether or not an unincubated egg is fertile. The yolk and the germ spot are enclosed by a thin membrane known as the viteline membrane.

The albumen or white of the egg is made up of alternate layers of thick and thin albumen. Surrounding the yolk is a thin layer of very dense albumen called the chalazeferous layer. From this layer of dense albumen extend the fibrous or cord-like structures called chalazae. These extend out towards the ends of the egg and are anchored in another layer of thick albumen. The chalazae tend to keep the yolk near the center of the egg.

Between the chalaziferous and outer layer of thick albumen there is an inner layer of thin albumen. If an egg is broken out and the thick albumen surface is cut, some of the thin albumen can be seen merging with the thick albumen.

After the egg is laid, a number of changes take place as far as the interior quality is concerned. The rate at which this takes place depends upon how the egg is handled. As mentioned earlier when an egg is laid there is no air cell, but as the contents cool and contract the air cell becomes apparent. A further increase in the size of the air cell is dependent upon the rate of evaporation of moisture from the egg. The rate of evaporation depends upon the thickness and texture of the shell as well as the temperature, relative humidity of the storage area and the length of the storage period. Air cell size is an important factor in grading eggs but actually has no significance as far as eating qualities are concerned.

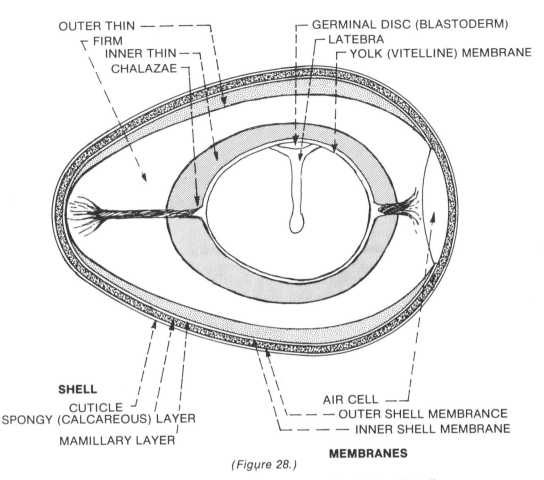

ALBUMEN

OUTER THIN
FIRM
INNER THIN
CHALAZAE

YOLK

GERMINAL DISC (BLASTODERM)
LATEBRA
YOLK (VITELLINE) MEMBRANE

SHELL

CUTICLE
SPONGY (CALCAREOUS) LAYER
MAMILLARY LAYER

AIR CELL
OUTER SHELL MEMBRANCE
INNER SHELL MEMBRANE

MEMBRANES

(Figure 28.)

During the holding period a change takes place in the thick albumen. It gradually breaks down into thin or watery albumen. If the egg is stored long enough, eventually there will be no evidence of thick albumen left. As a result of these changes, when eggs are broken out into a pan, the contents seem to spread over a large area and the yolk has a flattened and enlarged appearance. Most individuals frown on this type of egg because it is a sign of poor quality. An egg of high quality, when broken out, will have a large amount of thick albumen adhering to the yolk. The yolk will be upstanding and practically spherical in shape. When broken out, the egg will occupy a much smaller area in the pan. If a high quality egg that is hard-cooked is cut in two, the yolk will be well centered and the air cell small. With the low

145

quality egg the yolk may be close to, or touching the shell membrane and the air cell quite large.

Egg quality can be determined relatively easily with the use of a candling light. Candling is the process of twirling the egg before a light in a dark room. When a low quality egg is candled the yolk shadow will be plainly visible and off-center. The egg contents move about very rapidly when the egg is twirled. When an egg is of high quality the yolk appears centered, the shadow is only slightly visible, and the contents of the egg don't move as easily. Thus, candling is a fairly effective means of determining interior egg quality.

The standards of quality for individual shell eggs are set up by the United States Department of Agriculture *(Table 24).*

It should be noted that blood spots or meat spots are permissible in C grade eggs providing the single defect, or the total of several defects, is not more than ⅛″ in size. Eggs containing blood or meat spots, not within these tolerances, are classified as inedible eggs.

Blood spots in the egg are usually found on the surface of the yolk. They may vary in size from a small speck to a large clot with some of the blood diffused throughout the entire albumen. Blood spots are caused by the rupture of one or more small blood vessels in the yolk follicle at the time of ovulation or in the oviduct during egg formation.

Meat spots are either blood spots which have changed in color, due to chemical action, or tissue that has sloughed off the oviduct of the hen during egg formation. Possibly 2% of the eggs produced contain blood or meat spots.

In addition to the interior quality factors which we have discussed, there are also exterior quality factors which must be considered. The more important exterior egg quality factors include condition or soundness of the shell and cleanliness. Leakers, dented cracks, and eggs with rough or thin shells, can be detected during processing without candling. Blind checks or hairline cracks can be seen during candling. Dirty and stained eggs should not be packed if to be marketed.

Weight classes of shell eggs *(Table 25)* are set up by the United States Department of Agriculture. The system uses the weight in ounces per dozen. For example, a single egg weighing 2 ounces is called a 24 ounce egg because a dozen of this size egg weighs 24 ounces. Uniformity of size in a market pack of eggs is important and there should be no more than a 3 ounce variation between each dozen. Small individual egg scales or larger automated scales are available for sizing eggs.

TABLE 24

SUMMARY OF UNITED STATES STANDARDS FOR QUALITY OF INDIVIDUAL SHELL EGGS
(Specifications for Each Quality Factor

Quality Factor	AA Quality	A Quality	B Quality	C Quality
Shell	Clean	Clean	Clean to very slightly stained	Unbroken
	Unbroken	Unbroken	Unbroken, may be slightly ab-normal	May have slightly stained areas
	Practically Normal	Practically Normal		Moderate stains permitted of less than ¼ of shell surface
				Prominent stains or adhering dirt not permitted.
Air Cell	⅛ inch or less in depth	3/16 inch or less in depth	⅜ inch or less in depth	May be over ⅜ inch in depth
	May show un-limited move-ment, may be free or bubbly	May show un-limited move-ment, may be free or bubbly	May show un-limited move-ment, may be free or bubbly	May show un-limited move-ment, may be free and bubbly
White	Clear and Firm	Clear and rea-sonably firm	Clear, may be slightly weak or thin	May be weak and watery
				Small blood spots or clots may be present*
				*(aggregating not more than ⅛ inch in diameter)

Yolk	Outline only slightly defined	Outline may be fairly well defined	Outline may be well defined	Outline may be plainly visible
	Practically free from defects	Practically free from defects	May be slightly enlarged and flattened	May appear dark, enlarged and flattened before candling light.
			May show definite but not serious defects.	May show clearly visible germ development, but no blood due to such development. May show other defects which do not render the egg inedible.
				Small blood clots or spots (aggregating not more than 1/8 inch in dia.) may be present.

TABLE 25

UNITED STATES WEIGHT CLASSES FOR CONSUMER GRADES OF SHELL EGGS

Size or Weight Class	Minimum Net Weight Per Dozen	Minimum Net Weight per 30 Dozen Case	Minimum Weight for Individual Eggs at rate per Dozen
	Ounces	Pounds	Ounces
Jumbo	30	56	29
Extra Large	27	50 1/2	26
Large	24	45	23
Medium	21	39 1/2	20
Small	18	34	17
Peewee	15	28	—

Source: *Egg Grading Manual Agricultural Handbook No. 75*

care of eggs on the farm

As stated earlier, immediately after the egg is laid the quality begins to deteriorate. The sooner the egg is removed from the nest, cleaned, cooled and packed, the better it is. Some of the management recommendations that will lead to higher quality eggs are listed as follows:

1. Keep the birds confined.
2. Gather eggs frequently, at least three times a day.
3. If eggs must be cleaned, clean them immediately after gathering.
4. Dry clean slightly dirty eggs.
5. Cool eggs as quickly as possible.
6. Maintain an egg storage room temperature of 50-55° and a relative humidity of 75%.

cleaning eggs

If eggs are to be washed, they should be washed at a temperature of 110-115° F. in clean water, with an approved detergent sanitizer. Washing time should be no more than 3 minutes. If the eggs are rinsed following the washing, they should be rinsed with a water containing a detergent sanitizer and then dried and cooled. Slightly soiled eggs may be dry cleaned with an abrasive material such as emery cloth, fine sandpaper on a hand buffer, or steel wool.

On commercial egg farms the washing and sizing of eggs is done with large automated equipment. Smaller operations may use immersion washers. The eggs are immersed in warm detergent sanitizer which is agitated to clean the eggs.

Poultry Processing

At best the job of processing poultry is a messy one. Ideally there should be two rooms available for the processing procedure. One for killing and picking the birds, and the other for finishing, eviscerating and packaging. If this is not possible, then the killing and plucking should be done in one operation. The room should then be cleaned and the birds drawn and packaged as a second operation. This procedure will make the whole operation far more sanitary.

care before killing

Birds should be staved for about 12 hours before they are to be killed. This will give ample time for the crop and intestines to empty. Starving the birds makes the job of eviscerating much cleaner and easier. Birds that are to be starved should be removed from the pen and put into coops containing wire or slat bottoms, so that they do not gain access to feed, litter, feathers of manure.

Care should be taken in catching and handling the birds to prevent bruising. They should be caught by the shanks and not be permitted to flop their wings against equipment or other hard surfaces. This will help prevent bruising and poor dressed appearance. After the birds are caught, keep them in a comfortable, well ventilated place prior to killing. Overheating or lack of oxygen can cause poor bleeding resulting in bluish, discolored carcasses.

12

EQUIPMENT REQUIRED

The Shackles Or Killing Cones

When only a few birds are to be dressed, a shackle can be made from a cord with a block of wood 2" x 2" square, attached to the lower end *(Figure 29)*. A half-hitch is made around both legs and the bird suspended upside-down. The block of wood prevents the cord from pulling through. Commercial dressing plants use wire shackles which hold the legs apart and make for easier plucking. Some producers make their own shackles out of heavy-gauge wire *(Figure 30)*. Other people prefer to use killing cones which are similar to funnels. The bird is put down through the funnel with its head protruding through the lower end. This restrains the bird and prevents some of the struggling which often leads to bruising or broken bones.

the knives

About any type of knife is satisfactory for dressing poultry. There are specialized knives for killing, for boning, and for pinning. Six-inch boning knives work very well. If the birds are to be brained then a thin sticking knife should be used.

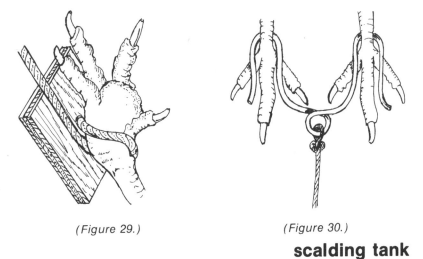

(Figure 29.)　　　　　　　　　*(Figure 30.)*

scalding tank

Where only a few birds are to be dressed, a 10-20 gallon garbage can, or any other clean container of suitable size, is satisfactory. Where considerable dressing is done a thermostatically controlled scalding vat is preferred.

thermometer

Accurate temperatures are important for certain types of scalding. You should have a good rugged dairy thermometer, or some other type of floating thermometer which accurately registers temperatures between 120-212° F.

weight or weighted blood cup

A weight or weighted blood cup attached to the lower beak of the bird will prevent it from struggling and splashing blood around. The weight may be made from a window weight and attached to the lower beak by means of a sharp hook. The blood cup is not used when killing funnels are available.

killing

The bird should be suspended by its feet with a shackle, or placed in a killing cone. The head is held with one hand, it is pulled down for slight tension to steady the bird. With a sharp knife the jugular vein is severed by cutting into the neck just back of the mandibles. This can be done by inserting the knife into the neck close to the neck bone turning the knife outward and severing the jugular. It may also be done by cutting from the outside. Another method which is sometimes used, is to cut the jugular vein from inside the mouth. Using this method the bird hangs with the breast toward the operator. The head is held firmly with the thumb and first finger at the ear lobes. A slight pull with pressure causes the beak to open. The knife is then inserted into the mouth, so that the point can be felt just back of the left ear lobe as you face it. It will be opposite for left-handers. With a slight pressure and drawing outward towards the opposite corner of the mouth the jugular vein is cut at the junction of the connecting vein running across the back of the throat *(Figure 31)*. Warning: the head should be held so that the fingers do not get in the way.

During the bleeding process the birds should be restrained by holding the head until the bleeding and flopping stops, or a weight is attached to the lower beak. The wings or legs should not be grasped tightly so as to restrict the flow of blood from these parts. A poor-appearing dressed carcass will result if the bleeding is incomplete.

(Figure 31.)

debraining

Debraining loosens the feathers so that it is easier to pluck the birds. It is done after the jugular vein is cut. Debraining is done when the birds are to be dry-picked, but may also be done when the birds are to be semi-scalded to make the removal of feathers easier.

The knife is inserted through the groove or cleft in the roof of the mouth and pushed through to the rear of the skull where it pierces the rear lobe of the brain *(Figure 31a)*. The knife is then given a quarter-turn. This kills the bird and loosens the feathers. A characteristic squawk and shudder indicates a good stick. If the front portions of the brain are pierced it may cause the feathers to tighten. This procedure requires considerable practice before the operator becomes proficient.

(Figure 31a.)

Picking

In general, there are four methods of removing feathers from birds. They are the hard scald, the sub-scald, the semi-scald and dry picking.

hard scald or full scald

This method is probably used more commonly on the farm than any others. The hard scald uses 160°-180° temperatures for 30-60 seconds. After the bird is sloshed up and down in water at this temperature, the feathers are removed very easily. With this method the time the bird is scalded depends upon the temperature of the water and the age of the birds. It should be only long enough so that the feathers can be pulled easily. This method would conceivably be used only for older birds and possibly waterfowl. Hard scalding makes for fast easy picking but destroys the protective covering of the skin. The hard scald causes a dark, crusty, blotchy, appearance and results in poorer keeping quality.

the sub-scald

The sub-scald uses a temperature of 138-140° F. for 30-75 seconds. This method causes a break-down of the outer layer of skin but the flesh is not affected as in hard scalding. The main advantage of the sub-scald is the easy removal of feathers and a uniform skin color. However, the skin surface tends to be moist and sticky and will discolor if not kept wet and covered. This method is frequently used for turkeys and waterfowl. The water temperature for scalding ducks is normally 135-145° F. and for geese from 145-155° F. The length of the scald is from 1½ to 3 minutes.

semi-scald

With the semi-scald the bird is sloshed up and down in water at a temperature of 125-130° F. Generally a water temperature of 128-130° for 30 seconds gives satisfactory results. The temperature and time varies with the age of the birds. Older birds require a higher temperature and a longer time.

Semi-scalding of poultry softens and spreads the fat beneath the surface of the skin. This improves the appearance of the dressed bird. Over-scalding is caused by too high a temperature for too long a time. This method is generally used for young birds such as broilers, fryers, roasters and capons.

One other technique may be utilized in connection with the semi-scald or the sub-scald, and this is known as wax picking. After the birds are scalded they are dipped in hot wax. After they are removed from the wax they are dipped in cold water to solidify the wax. The bird is coated with wax and the feathers and pin feathers become imbedded in it, so that when the wax is stripped from the carcass, the feathers and pinfeathers are easily pulled out. This method is commonly used in dressing waterfowl. It is not so frequently used when dressing chickens or turkeys.

A modification of the above described wax picking method is

sometimes used for dressing waterfowl. Parafin is added to the scald water to form a thick floating layer. The bird is held in the scalding vat for the required length of time and then removed slowly so that a layer of parafin adheres to the feathers. The parafin is given an opportunity to solidify and is then stripped off carrying with it many of the feathers and pin feathers which are normally hard to pull.

Waterfowl feathers tend to resist thorough wetting during the scalding process. A little detergent in the scald water will help solve this problem and make the job of picking somewhat easier.

pinning and singeing

Pinfeathers, the tiny immature feathers, are best removed under a slow stream of cold tap water. Use a slight pressure and a rubbing motion. Those that are difficult to get can be removed by using a pinning knife or dull knife. Applying pressure, the pinfeathers can be squeezed out. The most difficult may have to be pulled.

Chickens and turkeys usually have a few hair-like feathers left following the plucking operation. These hairs can be removed by singeing with an open flame. Singeing equipment is available commercially for large operations but for small units a small gas torch or a gas range work well. The birds are easily singed by rotating the defeathered bird in the flame.

Eviscerating

After the carcasses are picked and singed, they should be washed in clean, cool, water. As soon as they are washed they are ready to be eviscerated. Some prefer to cool the poultry before eviscerating and cutting it up, because it is somewhat easier and cleaner to do after it is cooled. Others eviscerate and then cool by placing the birds in ice water or cool water which is constantly replenished. There are many methods of drawing poultry. The following methods are not the only ones but are satisfactory. Neatness and cleanliness are essential if a satisfactory job is to be done.

The tools needed for drawing poultry are a sharp stiff-bladed knife, a hook if the leg tendons are to be pulled and a solid block or bench upon which to work. All equipment and working surfaces should be clean. A piece of heavy parchment paper, or meat paper may be laid on the working surface and changed as necessary.

eviscerating roasting chickens, turkeys and waterfowl

Roasters, turkeys and capons are usually stuffed and roasted so the drawing procedure is the same for all.

It is desirable when eviscerating turkeys to remove the tendons from the drumsticks before removing the shanks and feet. By cutting the skin along the shank, the tendons extending through the back of the leg may be exposed and twisted out with a hook or a special tendon puller, if one is available.

The shanks and feet should be cut off straight through or slightly below the hock joint, leaving a small flap of the skin on the back of the hock joint. This will help prevent the flesh on the drumstick from drawing up and exposing the bone during the roasting process. The oil sac on the back near the tail should be cut out, as it sometimes gives a peculiar flavor to the meat. This is removed with a wedge-shaped cut *(Figure 32)*.

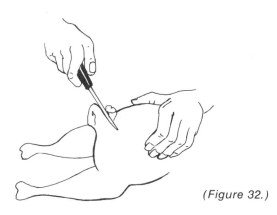

(Figure 32.)

To remove the crop, windpipe, gullet and neck, cut the head off, and slit the skin down the back of the neck. Separate the skin from the neck and then, from the gullet and windpipe. Follow the gullet to the crop and remove by cutting below the crop. The loose skin then serves as a flap which folds over the front opening and permits stuffing the bird without sewing. The neck is cut off as close to the shoulders as possible *(Figure 33)*. A pair of pruning shears are handy for this purpose. The flap of skin is then folded back between the shoulders and locked in place by folding the tips of the wings over it.

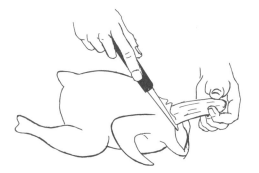

(Figure 33.)

The vent is loosened by cutting around it. This should be done carefully to avoid cutting into the intestine. The viscera are removed through a short horizontal cut approximately 1½-2 inches below the cut made around the vent. The horizontal cut should be about 3 inches long. *(Figure 34)*. The lungs, liver and heart attachments are carefully broken by inserting the fingers through the front opening. The intestines are loosened from the rear opeing by working the fingers around them and breaking the tissues that hold them in the body cavity. The viscera are removed through the rear opening in one mass by inserting two fingers through the rear opening and hooking them over the gizzard cupping the hand and using a gentle pull and slight twisting motion.

Remove the gonads (ovaries and testes). The gonads are attached to the backbone. They can be removed quite easily by hand. Surgical capons should not have gonads unless they are "slips".

The lungs are attached to the ribs on either side of the backbone. These can be removed by using the index finger to break the tissues attaching the lungs to the ribs. Merely insert the finger between the ribs, and scrape the lungs loose. The lungs are pink and spongy in appearance.

After all the organs have been removed, wash inside with a hose or under a faucet. Also wash the outside removing all adhering dirt, loose skin, pin feathers, blood or singed hairs. Hang the birds so as to drain the water from the body cavity. Twenty minutes should be adequate to thoroughly drain the birds.

(Figure 34.)

HORIZONTAL CUT

trussing

A properly trussed bird makes for a nicer appearance. It also conserves the juices and flavors during the roasting process. One method is to place the bird on its back and draw a string across the shoulders and over the wings. It is then crossed over the drumsticks, brought under the back, drawn tightly and tied at the base of the tail.

Another method of trussing is to place the hock joints under the strip or bar of skin between the vent opeining and the cut from which the viscera was removed. *(Figure 35).* The neck flap is drawn back between the shoulders and the wingtips are folded over the shoulders and serve to hold the skin.

The same procedures are used for killing and eviscerating waterfowl.

(Figure 35.)

cleaning the giblets

Remove the gall bladder — the green sac attached to the liver — without breaking it. If the gall bladder is broken while removing the viscera, or cleaning the liver, it will likely give a bitter, unpleasant taste to any part with which it comes in contact as well as causing a green discoloration.

If the gizzard is cool and care is used, it may be cleaned without breaking the inner lining. Cut carefully through the thick muscle until a light streak is observed (do not cut into the inner sac or the inner lining). The gizzard muscle may then be pulled apart with the thumbs and the sac, and its contents removed unbroken.

eviscerating broilers and fryers

The method used for drawing broilers and fryers is very similar, except that fryers are usually cut up into more pieces. Both can be drawn the same as roasters but the usual method is to split and cut up the fryers.

The first operation is to cut off the legs at the hock joint. Remove the oil sac at the base of the tail and cut off the neck. Place the bird on its side, neck toward the operator, and the back of the bird toward the cutting hand. Split the bird with a single cut down the backbone or, if preferred, two cuts can be made along either side of the backbone and the backbone and neck stripped out. The cuts along the backbone should be made with a stiff bladed knife and just deep enough to sever the ribs. The vent is cut out, and the viscera then removed from the bird.

To remove the breast bone, simply make a small cut in the cartilage toward the front of the breast bone. By applying a downward pressure on both sides of the bone with the thumbs and snapping it, the breast bone can be pulled out. Broilers are usually left whole or in halves and fryers, quartered.

eviscerating fowl

Fowl or old laying hens may be drawn similar to roasters and turkeys. However, they are too tough for barbecuing, roasting or frying and are usually cut up for stewing or fricasse. One method of cutting-up fowl is as follows: Remove the head, neck, crop, gullet and windpipe as described under roasting chickens, turkeys and capons. Next remove the feet, oil sac, wings and legs. A slight cut is made on both sides of the body of the bird to the rear and below the breast, cutting to the backbone. The fowl is then broken apart by bending back the breast. The breast, back and neck, and liver are then cut away leaving the viscera in one mass. The breast may be split into halves if desired. The drumsticks and thighs are separated by cutting through the second joint above the hock joint. The giblets are cleaned as described earlier.

chilling and packaging

It is important that the body heat be removed from the birds as soon as possible after killing. If the cooling is done slowly, bacteria can develop, causing spoilage and undesirable flavors.

If birds are to be air-dried, the air temperature should be from 30-35° F. to cool the birds properly. The time required to cool the carcasses depends on the size of the birds and the temperature of the air. It may vary from 6½ hours for a 3 pound fowl to 10 hours for a 7 pound male bird.

Water cooling of poultry can be used when air temperatures are above 35° F. If the birds have been dressed in scalding temperatures that are too high, or for too long a period of time, the air-dried birds may show a blotchy, discolored appearance of the skin. When the hard scald is used water cooling is the preferred method of cooling the carcasses. Dressed birds should be cooled in pails or tanks of ice water or cold running water. The important factor is to maintain a constant temperature of 34-40° F. To cool the birds to an internal temperature of 36-40° F. requires 5-10 hours in the water, depending again, upon the size of the carcass.

In the case of cut-up poultry, the birds may be cut up immediately after dressing, thrown into cold water and cooled in a much shorter time. It should again be noted that it is much easier and cleaner to draw or cut up poultry after it has been cooled.

Birds should be cooled and aged for approximately 8-10 hours. If eaten or frozen immediately after dressing, the carcasses will be tougher than if aged for a period of time.

Remove the carcasses from the running water or ice water and hang them up to dry for 10-30 minutes before packaging. Every effort should be made to get all of the water out of the body cavity of the bird before putting it into the bag.

The giblets should be wrapped in a square sheet of wax paper or a sandwich size plastic bag. Giblets can be stuffed into the body cavity. The giblets should be wrapped well, so that if they spoil the carcass will not be affected.

There are two types of bags available for poultry — one is the common plastic bag; the other is the so called cryo-vac bag. The cryo-vac bag is a shrinkable bag and one that adheres closely to the bird after shrinking in boiling water. It makes a much nicer appearing package. It also helps to reduce the amount of water loss during the freezing process. However, good plastic bags are available and will do a satisfactory job of maintaining quality in frozen dressed poultry. The bags should be highly impermeable to moisture to prevent dehydration in freezer storage, which causes toughness.

Birds to be bagged should be thoroughly trussed and then inserted front end first into the plastic bag. The bag should be excessively large. After the bird is in the bag, the excess air can be removed by applying a vacuum cleaner or by inserting a flexible hose into the top of the bag and applying a vacuum. Merely keep the bag snug around the hose or vacuum cleaner then suck the air out of the bag. Twist the bag several times and secure it with a wire-tie or rubber band.

Fresh-dressed, ready-to-cook poultry has a shelf life of approximately 5 days. It should be refrigerated at a temperature of 29-34° F. If it is to be frozen, it should be frozen by the third day after it is dressed and chilled. Poultry to be frozen should be chilled to below 40° F. before placing in the freezer.

The weight losses from live to dressed weight vary with the type, age and size of bird *(Table 26)*.

TABLE 26

PROCESSING YIELDS — POULTRY

Type	Approximate % Yield Live-to-Eviscerated
Broilers and Fryers	75
Roasters	76
Fowl — Leghorn-Type	68
Fowl — Heavy-Type	70
Turkeys — Heavy-Roasters	80
Turkeys — Light-Roasters	78
Ducks	70
Geese	68-73

Source: *New England Poultry Management and Business Analysis Manual*

state and federal grading and inspection

Processors of poultry and eggs that are to be sold off the farm are subject to the Poultry Products Inspection Act and the Egg Products Inspection Act. There are exemptions for small producers. Check into them if you have questions. For information on grading and inspection programs and how they affect you, contact your State Department of Agriculture.

13

FLOCK HEALTH

D isease is not normally much of a problem with small flocks, but there are several diseases that can possibly affect your flock. It was emphasized earlier that an ounce of prevention is worth a pound of cure. If you purchase stock from a good clean source, follow a sound sanitation program, use a good feeding program and provide a comfortable house with ample dry litter and plenty of fresh air, you will have gone a long way toward keeping your flock healthy.

It should be realized that losses will occur in any flock. Commercial flock owners expect a monthly mortality of approximately ¾ to 1%. However, a disease in the flock is usually manifested by a drop in feed and water consumption, a drop in egg production and sick or dead birds. When it is apparent that a disease is present in the flock, seek the advice of a person who is a trained poultry diagnostician. It is not advisable to use drugs or antibiotics indiscriminately. Sometimes this will do more harm than good and the only result may be a waste of money.

For the list of diagnosticians in your area consult the list at the back of this text. When birds are submitted to a State Diagnostic Laboratory, a sample of the flock problem should be submitted. The sample should include two or more sick birds, or recently dead birds. Care should be taken to preserve dead specimens by cooling or freezing so as to prevent decomposition. An early diagnosis and fast treatment is always recommended for a quick solution to poultry disease problems.

There are many poultry diseases but only the more common ones will be discussed in this text. These will not be discussed in great detail because this text is intended mainly for the small flock situations. There are many excellent texts on poultry diseases if more in-depth information is desired.

Many of the diseases which will be described are common to both chickens and turkeys. To avoid duplication, these will be covered in the one section. Those diseases that are peculiar to or more important in waterfowl will be covered in the section on Diseases of Waterfowl.

Diseases
aspergillosis or brooder pneumonia

Aspergillosis is a disease of young birds including chicks, turkeys and ducks. Symptoms include, dumpy-acting birds, rapid breathing, gasping, and possibly inflamed eyes.

The disease is caused by a fungus which is inhaled by the birds, usually from moldy litter or feed. On post mortem, yellowish-green nodules may be found on the lungs, in the trachea, bronchi, and viscera. There is no treatment known. However, further spread of the infection may be prevented by culling the sick birds and cleaning and disinfecting the house and equipment thoroughly. Musty litter or feed must be carefully removed from the building to prevent further spread of the disease.

blackhead or histomoniasis

Blackhead is caused by a protozoan parasite. It is infrequently found in chickens but is a very common disease of turkeys of all ages. Since it can affect both chickens and turkeys, and chickens may act as an intermediate host for the organism causing blackhead, it is commonly felt to be rather risky to keep chickens and turkeys on the same farm. The term blackhead is misleading, because this symptom may or may not be present.

Cecal worm eggs can harbor the organism over long periods of time and when picked up by the turkeys infect the intestines and liver. Mortality may reach 50% if treatment is not started and the infection checked immediately. Symptoms may be droopiness and dark heads, brownish colored and foamy droppings. Inflamation of the intestine and ulcers on the liver may be seen upon autopsy. The incidence of the disease and its severity is dependent upon the management and sanitation programs used. Management factors such as sanitation of the brooding facilities, rotation of the range areas, and segregating young birds from old birds help to prevent outbreaks. Segregation of turkeys from chicken flocks is very helpful in preventing problems with blackhead.

Blackhead drugs are commonly added to the feed to prevent outbreaks and may be used to treat blackhead infections.

bluecomb

The cause of bluecomb is not known. It affects both chickens and turkeys, it can hit birds at any age but usually affects chickens and pullets early in the laying cycle. Mortality can be heavy. When an outbreak occurs in laying chickens or laying turkeys, egg production is severely affected.

Characteristic symptoms include a loss of appetite, darkening of the head and a watery diarrhea. The birds get listless and drink excessive amounts of water.

Antibiotics given in the drinking water usually result in a dramatic recovery when treatment is started early.

bumblefoot

Bumblefoot is the swelling of the foot involving the foot pad and the area around the base of the toes. It is thought to be caused by injuries resulting from bruising, a penetration of wire or thorns or other sharp objects with which the feet come in contact. Organisms such as E. Coli, Streptococcus and Staphylococcus have been isolated from the puss material found in the swollen area.

Equipment should be designed so that the birds do not get foot punctures from wire or nails. A deep dry litter is helpful in preventing this condition, especially in those areas in front of high roosts or nests which require that the birds jump to the floor.

Birds with severely swollen feet, or birds that are very lame should be removed from the flock. Some birds may recover if the scab over the swollen area is removed and the core is squeezed out. The foot should then be washed with an antiseptic.

fowl cholera

Fowl cholera is caused by a bacterium. It is a highly infectious disease of all domestic birds including chickens, turkeys and ducks.

The birds become sick rapidly and may die suddenly without showing external symptoms. They may appear listless, feverish, drink excessive amounts of water and show a diarrhea.

Post mortem findings include red spots or hemorrhages on the surface of the heart, lungs, intestines or in the fatty tissues. The birds may have swollen livers (a cooked appearance) with white spots. Treatment with sulfonamides such as sulfaquinoxalene, sulfamethazine and others are currently recommended. Sulfaquinoxalene in the feed at the .33% level for 14 days is considered to be one of the best treatments. Antibiotics are sometimes injected at high levels.

Sanitation in the poultry house, range rotation and proper disposal of dead birds help to prevent cholera outbreaks. In problem areas, birds may need to be vaccinated.

chronic respiratory disease:
or mycoplasma gallisepticum

Chronic respiratory disease is an infectious and contagious disease affecting birds of all ages. It is prevelent in all areas of the country. It is caused by a pleuro-pneumonia-like organism. It affects chickens, turkeys and ducks.

C.R.D. is a respiratory disease affecting the whole respiratory system, especially the accessory lungs or air sacs. It is not a killer but can cause considerable morbidity, especially if treatment is not undertaken soon and secondary invaders avoided.

In young birds the signs are rattling, sneezing and sniffling; In older birds, the condition may go unnoticed. When the condition becomes complicated, there may be a nasal discharge, a foamy exudate from the eyes. There may be difficult breathing, lack of appetite and a drop in egg production in producing birds. Young birds may be somewhat stunted and unthrifty.

Post mortem findings include, cloudy air sacs with cheesy exudates, mucus in the nasal passages, trachea and lungs.

Transmission is through the hatching egg, through the air and via infected equipment, feed bags, and other means.

Several of the antibiotics are used to treat the disease and may be administered in the drinking water, the feed or by injection. These currently include tetracycline, tylosin and spectinomycin and others.

Stress appears to play a key role in triggering the disease. Thus, good environmental conditions and good management are important.

infectious bronchitis

Infectious bronchitis is caused by a virus. It spreads very rapidly and hits birds of all ages and is found in chickens. It may cause up to 50% mortality in young chicks and 5-10% in adult birds. It affects the reproductive organs of young birds thus affecting future performance.

Symptoms include sneezing and coughing. Egg production drops to nearly zero in laying birds and the eggs become misshapened, soft shelled, chalky or porous and light in color in the case of brown eggs. The interior egg quality is also affected.

Post mortem findings may be inflamed nasal passages with cheesy plugs in the lower trachea and bronchi. There is no known treatment for infectious bronchitis. Treatment with antibiotics for 3 to 5 days is usually recommended to ward off secondary invaders.

Control of infectious bronchitis is accomplished through a vaccination program. Vaccines may be administered by mass methods or individually by nasal or ocular drops. Mass methods of vaccination include drinking water, dust, or spray. The manufacturer's recommendations should be followed in the use of any of these products. Bronchitis vaccine is frequently given in combination with Newcastle vaccine.

coccidiosis

Coccidiosis is primarily a problem in chickens and turkeys, though on rare occasions it may be found in flocks of ducks. Coccidiosis is a very common disease of poultry and is caused by a protozoan parasite, coccidia. The birds expose themselves to the disease by picking up sporulated oocysts in fecal material and litter. It should be assumed that all flocks grown on litter are exposed to the disease. Birds grown on wire are not exposed to droppings and normally don't become exposed to coccidiosis.

Coccidia are host specific, that is, the coccidia that affect chickens do not affect turkeys. About 9 species of coccidia affect chickens. Of these nine, three cause most of the problems. Different species affect various parts of the digestive tract. Six species are known to infect turkeys, but only two of these commonly cause problems.

Symptoms of coccidiosis are ruffled feathers, unthriftiness, head drawn back into the shoulders, an appearance of being chilled and a diarrhea which may be bloody with some forms. If permitted to go unchecked, considerable mortality, morbidity, and poor flock results may occur.

On post mortem, lesions and hemorrhages may be seen in the various parts of the intestine, depending upon the species involved.

The disease may be prevented by feeding coccidiostats in the growing diet and permitting the bird to build an immunity or completely controlling by feeding a preventative level of coccidiostats.

Treatment involves the use of sulfonamids or other coccidiostats as prescribed by a diagnostician or service man.

Infectious Coryza

Infectious coryza is a respiratory disease of chickens caused by a bacterium. It spreads rapidly from bird to bird and is thought to be triggered by a stress. Mortality may be high but is usually low. Symptoms include an involvement of the sinuses and nasal passages with a nasal discharge and swelling of the face. Egg production drops in laying flocks.

Post mortem findings usually include congested nasal passages and fluid-filled tissues of the face and wattles.

Old birds are carriers and should not be mixed in with young birds, but either sold or kept isolated from growing flocks. Sulfonamids are frequently used for treating Coryza.

erysipelas

Erysipelas is primarily a disease of turkeys. It may affect ducks and rarely affects chickens. It is caused by a bacterium. Swine, sheep and man are also susceptible.

Symptoms are swollen snoods, bluish purple areas on the skin, and congestion of the liver and spleen. Birds may become listless, have swollen joints and exhibit a yellowish-green diarrhea. It is primarily a disease of toms because of injuries during fighting. The erysipelas organism readily enters through skin breaks.

Erysipelas is considered to be a soil-borne disease and contaminated premises are assumed to be the primary source of infection.

The disease responds to penicillin. Tetracycline is also effective. Control requires good management and sanitation.

fowl pox

Fowl pox is found in many areas of the country. It is caused by a bacterium and spread by contact with infected birds or by such vectors as flies and mosquitos or wild birds. There are two forms of fowl pox. The dry or skin-type and the wet or throat-type. The same organizm causes both. The disease affects both chickens and turkeys.

Birds with fowl pox have poor appetites and look sick. The wet pox causes difficult breathing, a nasal or eye discharge, yellowish, soft cankers on the mouth and tongue. With the dry pox, the birds develop small greyish white lumps on the face, comb and wattles. These eventually turn dark brown and become scabs.

On post mortem, cankers may be found in the membranes of the mouth, throat and windpipe. There may be occasional lung involvement or cloudy air sacs.

There is no treatment for the disease itself, though an antibiotic may help to reduce the stress of the disease.

The only means of control is through administering a vaccine. This is recommended in those areas where fowl pox is a problem.

laryngotracheitis

Laryngotracheitis is a viral disease. It spreads very rapidly. It may be airborne, spread by contact or by carrier birds. It affects all ages of chickens and pheasants. Its affect on birds under four weeks is mild, but may be very severe in adult birds. Mortality in some cases may be as high as 60%. It causes egg production to drop but has no affect on egg quality.

Signs of the disease include coughing and sneezing and, as the disease progresses, the birds find it difficult to breath. The neck is outstretched when the birds inhale and this may be accompanied by a cawing sound. A bloody mucous may be expelled when the birds cough and sneeze.

On post mortem examination the throat, trachea and larynx are very inflamed and swollen. There is frequently a bloody mucus or yellowish, false membrane in the trachea. Severe inflammation of the larynx and trachea are in evidence and occasional cheesy plugs are found in the trachea. No treatment for laryngotracheitis is effective. In those areas where the disease is a problem, a vaccination program should be followed.

After an outbreak of laryngotracheitis, some birds are carriers. These birds are a possible source of infection for non-immune birds.

lymphoid leukosis

This disease has caused serious economic losses to the poultry industry for many years. Most forms of lymphoid leukosis occur in the older birds or laying flocks. Mortality from this disease is seldom a problem in the young. Infected birds begin to develop tumors of the internal organs, particularly the liver, and slowly lose weight and eventually die. Acute death losses are not common with lymphoid leukosis but occur a few at a time over a long period of time.

There is no known treatment for the disease and it remains a perplexing problem for the poultry industry. Although isolation brooding, and clean brooding facilities are recommended, these management practices have proved to be ineffective in controlling lymphoid leukosis.

marek's disease

Marek's disease is a disease of chickens caused by a herpes virus. The disease takes many forms and affects many areas of the body, including the muscle, nerves, skin and intestines. Birds over two weeks of age may be affected. It can cause losses of 30% or more in pullet flocks up to the time of housing.

There is no treatment for Marek's Disease and the best control measure is by vaccination of the baby chicks. This is usually done at the hatchery.

newcastle disease

Newcastle disease is widespread. It is acute, highly contagious, and found in chickens, turkeys, ducks and geese. It is a respiratory disease caused by a virus. It causes high mortality in young flocks. Production in laying birds frequently drops to zero.

Newcastle spreads rapidly through the flock causing gasping, coughing and hoarse chirping. Water consumption increases and a loss of appetite occurs. Infected birds tend to huddle and exhibit signs of partial or complete paralysis of the legs and wings. The head may be held between the legs or on the back with the neck twisted.

The disease is transmitted in many ways. It can be tracked in by people, brought in with chickens from another premise or on dirty equipment, feed bags or by wild birds which gain entrance to the pen.

Post mortem findings may include congestion and hemorrhages in the gizzard, intestine, and proventriculas. The air sacs may be cloudy.

There is no effective treatment, though antibiotics are normally given to keep down secondary invaders. Vaccination is recommended in most areas of the country, and can be administered individually or on a mass basis. Recommendations vary with the area and the type of bird. Newcastle disease vaccines can be administered intranasally, ocularly, or by wing-web. On a mass basis it can be given alone or in combination with infectious bronchitis — in the drinking water, or in the form of a dust or spray. Manufacturers' recommendations should be followed for use of the product. The vaccination program should be that which is recommended for the area and the type of bird that is to be grown.

omphalitis

Omphalitis is caused by a bacterial infection of the navel. It occurs when the navel doesn't close properly following hatching. It is found in young chicks and turkey poults and may be caused by poor incubator or hatchery sanitation, chilling or overheating.

Birds with omphalitis are weak and unthrifty and tend to huddle together. The abdomen may be enlarged and feel soft and mushy. The navel is infected. The area around the navel may be a bluish-black in color. Mortality may be high for the first four to five days.

There is no treatment for the disease. Most of the affected chicks die the first few days. No medication is needed for the survivors.

paratyphoid

Paratyphoid is an infectious disease of chickens, turkeys, ducks and other birds and animals. It is caused by one or more of the salmonella bacteria. There are several types but Salmonella typhimurium is one of the most common in poultry and it accounts for over half of the typhoid outbreaks in poultry flocks. Transmission may be from the hen through the egg and to the chick. The organism is also found in fecal material of infected birds.

The disease is primarily one of young birds, but older birds may also be affected. In young birds mortality may run as high as 50%. Some birds may die without showing symptoms, while others show signs of weakness, loss of appetite, diarrhea and pasted vents. Birds may appear chilled and huddle together for warmth. In older birds there is a loss of weight, weakness, loss of egg production and diarrhea.

On post mortem of young birds, there may be unabsorbed yolk sacs, small white areas on the liver, inflammation of the intestinal tract, congestion of the lungs, and enlarged livers. Older birds usually show no lesions, though a few may show white areas on the liver.

Sulfamerazine, nitrofurans and some of the antibiotics may reduce losses, prevent secondary invaders, and increase appetite.

Control is through sanitation and isolation of the flock from sources of infection such as wild birds, birds from other flocks, contaminated feed and equipment.

pullorum

Pullorum is an infectious disease of chickens, turkeys and some other species. It is found all over the world. Pullorum is highly fatal to birds under two weeks of age. It is caused by a bacterium, salmonella pullorum.

Mortality begins at 5-7 days of age. The birds appear droopy, huddle together, act chilled and may show diarrhea and pasting of the vent. It is sometimes called white diarrhea.

Salmonella pullorum is spread mainly from the hen to the chick through the egg. It spreads rapidly through transmission on chick down in incubators and hatchers.

Post mortem findings in infected chicks include dead tissue in the heart, liver, lungs and other organs and an unabsorbed yolk sac. In adult birds there are discolored and misshapen ova; the heart muscle may be enlarged and show greyish-white nodules. The liver may also be enlarged, yellowish-green and coated with exudate. One of several types of blood tests help to establish a positive diagnosis for salmonella pullorum..

Sulfamethazine and sulfamerazine are effective in reducing mortality. Nitrofurans may also be effective in reducing mortality but will not cure the disease. Medication may hinder diagnosis.

Flocks which have recovered from salmonella pullorum should not be kept for replacements or breeding purposes unless they have been blood tested and found to be free of the disease.

infectious sinusitis: mycoplasma gallisepticum

Infectious sinusitis is a disease of turkeys and is caused by the same organism that causes chronic respiratory disease (CRD) in chickens.

It is transmitted through the eggs from carrier hens. As in the case of CRD, stress is thought to lower the poults resistance to the disease.

Symptoms are nasal discharge, coughing, difficult breathing, foamy secretion in the eyes and swollen sinuses, accompanied by a drop in feed consumption and loss of body weight. Air sac involvement may be in evidence on post mortem.

Antibiotics and antibiotic-vitamin mixtures will help. Individual treatment with injectible penicillin and streptomycin in the sinuses is recommended.

infectious synovitis: mycoplasma synoviae

Synovitis is an infectious disease of chickens and turkeys. It is caused by a pleuro-pneumonia-like organism (PPLO). At first it was identified as a cause of infections of the joints, but more recently it has been identified with a respiratory disease. The disease can affect birds of all ages. It is found in all areas of the United States and is a serious problem, particularly in broiler flocks.

Symptoms include lameness, hesitancy to move, swollen joints and foot pads, loss of weight and breast blisters. Some flocks show signs of the respiratory problem. Dying birds may show a greenish diarrhea.

Upon post mortem examination, swelling of the joints and a yellow exudate, especially in the hock, wing and foot joints, is in evidence. Internally, the birds may show signs of dehydration and enlarged livers and spleens. Birds showing the respiratory involvement are not easy to spot. On post mortem examination, the air sacs may be filled with liquid exudates.

The most common means of transmission is through infected breeders. Poor sanitation and management practices also contribute. Treatment is prescribed for market egg flocks. Breeder flocks with infectious synovitis should probably be disposed of. It can be passed on to their offspring. Antibiotics yield some results. These should be given by injection or in the drinking water. Some prefer to use both simultaneously for best results.

fowl typhoid

Fowl typhoid is caused by a bacterium, salmonella gallinarum. It affects chickens, turkeys, ducks and other species of birds. The disease may be present wherever poultry is grown.

Affected birds may look ruffled, droopy and unthrifty. Other symptoms may include pale combs and wattles, loss of appetite, increased thirst and a yellowish-green diarrhea.

On post mortem, the liver may have a mahogany color. The spleen may be enlarged and there may be pinpoint necrosis in the liver and other organs. There may be pinpoint hemorrhages in the fat and muscle tissue and evidence of enteritis.

Prevention, treatment and control are the same as for pullorum disease.

Diseases Of Waterfowl

Waterfowl raised in small flocks very seldom have disease problems. Where waterfowl are kept in large numbers, and concentrated in relatively small areas, disease may be more prevalent. Some breeds of waterfowl exhibit more resistance to diseases than others. For example, the Muscovy duck appears to be more resistant to diseases common to the Pekin and Runner.

Only the more common diseases of waterfowl will be considered in this section. It should again be emphasized that when disease problems occur it is well to get a reliable diagnosis from a poultry diagnostician.

amyloidosis

This is one of the common diseases of adult ducks and geese. Amyloidosis is also called wooden liver disease and can be quite readily recognized by the hardness of the liver. Mortality may reach as high as 10% in some flocks. The cause of the disease is not known and there is no known cure. Upon post mortem there are frequently large accumulations of fluids within the body cavity.

aspergillosis or brooder pneumonia

This disease was discussed earlier under chicken and turkey diseases. The causes and control measures are the same for waterfowl.

botulism

Botulism occurs in both young and adult waterfowl and in other types of poultry. It is caused by the bacteria, clostridium botulinum. This organism grows in decaying plant and animal material. Birds feeding on material containing the toxins produced by the bacteria lose control of their neck muscles. For this reason it is sometimes referred to as "limber neck". Infected waterfowl may drown if swimming water is available to them. The maintenance of dry, clean, facilities, the removal of decaying vegetation, or spoiled feed, will usually prevent this disease. A mixture of epsom salts, at the rate of one pound per 5 gallons of water, has been reported as an effective treatment.

fowl cholera

Fowl cholera has been discussed earlier under chicken and turkey diseases. The disease also attacks ducks and geese. The control and treatment for fowl cholera in waterfowl is the same as for other birds.

coccidiosis

As mentioned earlier, coccidiosis is a very infrequent problem with ducks. The species causing the disease in ducks is different from those causing the problem in chickens and turkeys. Control and treatment is the same as for treatment of other species of birds.

duck virus enteritis or duck plague

This disease affects ducks, geese, swans and other aquatic birds. It is caused by a virus that is transmissible by contact. It commonly occurs where water is available for swimming. The disease has affected ducks in commercial flocks in the eastern United States.

Signs of the disease are watery diarrhea, a nasal discharge and general droopiness. The symptoms develop 3-7 days after exposure. Death frequently results in 3 or 4 days. Generalized hemorrhages in the body organs may be seen on post mortem.

There is no satisfactory treatment. Only strict sanitation and the rearing of ducks in pens with access to drinking water will greatly reduce and help control the disease.

The exposure of domestic ducks to duck virus enteritis from migratory waterfowl can be avoided by keeping the susceptible ducks in houses or in wire enclosures.

keel disease

Keel disease occurs in young ducklings the first few days after hatching. Affected birds appear thin and dehydrated.

Fumigation of the hatching eggs, and a thorough washing and fumigation of the hatcher between hatches, are recommended to reduce the number of bacteria to which the young ducklings are exposed. Clean, warm, brooding facilities and good feed and fresh water will also help control the disease.

new duck disease or infectious serositis

New Duck disease is one of the most serious diseases affecting ducklings. It is caused by a bacterium. The symptoms resemble those of chronic respiratory disease of chickens. The first signs of the disease are sneezing and loss of balance. Losses up to 75% have been recorded. Death often is due to water starvation, rather than to the primary infection. Antibiotics and sulfa drugs have been used with some success.

virus hepatitis

Virus hepatitis outbreaks can cause 80-90% mortality in flocks of young ducklings. It is a highly contagious disease and strikes ducklings from 1-5 weeks of age.

Vaccination of the female breeding stock will prevent this disease. Antibodies produced by the vaccinated laying ducks are passed on through the egg to the young ducklings.

Parasites Of Poultry

Although there are many parasites of poultry, relatively few of them are of major importance. Some of the parasites live inside the bird, while others live on the outside. Certain of the external parasites affect growth and egg production. Some of the internal parasites can cause setbacks in weight gains, loss of egg production and even deaths if the infestation is severe. A few of the intestinal parasites harbor other disease organisms which are harmful to chickens and turkeys.

Internal Parasites
large roundworm

A light infestation of roundworms probably does little damage to the bird. However, when worms become numerous the birds can become unthrifty and feed conversion can suffer. The worms themselves seldom cause mortality but when present with other diseases may cause an increase in mortality.

The large roundworm is 1½″ to 3″ long. It is found in the middle of the small intestine and causes setbacks in weight gains and a loss of egg production in adult birds. The piperazine compounds are used to treat roundworms, and can be given in the water, in the feed, or in capsule form.

Birds infect themselves with the roundworm by picking up the egg from feces, the litter or other infested materials. It is, therefore, desirable to keep the litter as clean as possible to prevent the birds from picking up the eggs. Birds on wire floors or in cages do not have worms.

capillaria worm

The capillaria worm is a very small worm ½″ to 1½″ in length. It inhabits the upper part of the intestine, including the esophagus, crop and occasionally the ceca. It can cause inflammation of the intestines, diarrhea, weakness and, in laying birds, lowered egg production. It is very difficult to see capillaria worms with the naked eye, and so sometimes their presence is overlooked on autopsy. Capillaria worm infestations prevent the efficient utilization of Vitamin A in the bird. Frequently the treatment rations are supplemented with additional Vitamin A. Hygromycin added to the diet has been reported to be helpful in some cases. Capillaria has the same direct life cycle as the roundworm. The embryonated egg is picked up from the litter by the bird. The earthworm is also an intermediate host. Sanitary conditions are very important for the control of capillaria worms. It is very difficult to get rid of them where dirt floors are used in poultry buildings.

cecal worm

The life cycle of the cecal worm is similar to that of the large roundworm. These worms are found in the ceca instead of the small intestine. Their length is ½″ to ¾″. By itself it probably has little economic significance in chickens, but its eggs harbor the organism which causes blackhead disease of turkeys. In severe infestations, cecal worms may contribute to a somewhat unthrifty condition in the birds, weakness and emaciation. It is rather hard to treat because most of the intestinal contents bypass the ceca during the process of digestion. Phenothiazine or hygromycin B have been used successfully as treatments.

gizzard worm

The gizzard worm is reddish in color about ½"-1" in length. It is found under the lining of the gizzard and it impairs digestion due to the damage to the gizzard. No treatment is known, but the disease may be controlled by avoiding exposure of the birds to such intermediate hosts as grasshoppers, beetles, weevils and other insects.

gape worm

The gape worm attacks the bronchi, trachea and lungs. It can cause pneumonia, gasping for breath or even suffocation. High mortality may occur in young birds, especially turkeys and pheasants. The treatment of choice is to dust with barium antimonal tartrate, given at the rate of one ounce per 8 cubic feet of pen space for fifteen minutes. The earthworm is an intermediate host. The gape worm is y-shaped and red in color.

tape worm

There are several species of tape worms varying in size from microscopic to 6-7" long. They are flat, white and segmented. They inhabit the small intestine and cause a loss of weight, and a loss of egg production in adult laying birds. Treatment recommended for tape worms is dibutyltin dilaurate. This drug is sold in combination with piperazine and is effective in the removal of tape worms. One caution should be observed in the use of this material. It should be used only for severely infected flocks and should not be fed to laying birds.

The effective control of internal parasites depends primarily upon a program of cleanliness and sanitation. Parasitic eggs can remain viable in the soil for more than a year. This makes it important for producers to practice rotation of poultry runs or yards and if possible, to keep them clean and under cultivation to prevent the birds from picking up the parasites.

External Parasites

lice

Lice are chewing or biting insects that cause birds considerable grief. With severe infestations, egg production, growth and feed efficiency can suffer. They cause an irritation of the skin with a scab formation.

Lice spend their entire lives on the birds. They will die in a matter of a few hours if they get off the bird. The eggs are laid on the feathers where they are held with a glue-like substance. The eggs hatch in a few days to 2 weeks. Lice live on the scales of the skin and feathers. There are several types of lice which attack poultry. They include the body louse, shaft louse, wing louse, large chicken louse and the brown chicken louse. The body louse is one of the most common lice of poultry and usually affects older birds. The lice and its nits (eggs) are seen on the fluff, the breast, under the wings and on the back. The head louse attacks the feathers of the head and seems to be most harmful to young birds.

Materials which may currently be used to treat lice are carbaryl, malathion, or co-ral. These should be used according to the directions on the label. The birds should be examined frequently for signs of a reinfestation.

Mites

northern fowl mite

Northern fowl mites are a reddish dark brown in color. They are found around the vent, the tail and the breast of the birds. These mites live on the birds at all times. They may be seen on eggs in the nest when there is a severe infestation. They attack the feathers, suck blood, cause anemia, loss of weight and loss of egg production. Materials recommended for treatment of northern fowl mite are carbaryl or co-ral. Use according to the manufacturer's directions.

red mites

Red mites are not found on the birds during the daytime. Red mites feed during the night time. They may be seen on the underside of roosts, cracks of the wall or seams of the nest during the daylight hours. Other signs are salt and pepper-like trails under roost perches or clumps of manure on the furniture. Red mites are blood suckers. They cause irritation, loss of weight, loss of egg production, and anemia. Laying birds may refuse to use infested nests. The red mite may be treated with the same materials as those used for northern fowl mite. Use according to the manufacturer's recommendations.

depluming mite

The depluming mite burrows into the skin and causes an irritation at the base of the feather. In an attempt to relieve the itching, the bird pulls out feathers until they may be nearly naked in cases of severe infestation.

To control the depluming mite, use a dip containing 2 ounces of sulfur and 1 ounce of soap per gallon of water. The birds should be dipped so as to wet the feathers to the skin. Dip only on warm days. The treatment should be repeated in 3 or 4 weeks if necessary.

the tropical chicken flea

This flea clusters on the comb, wattles and earlobes and around the eyes of the bird. It causes irritation to the membranes of the eye and may cause eventual blindness and death. It causes a loss of weight and lowered egg production. Use carbaryl (sevin) in dust or spray form according to the recommendations on the label.

fowl tick or blue bug

The fowl tick is a flat, egg-shaped, reddish-brown bug. It resembles the ticks. The adults are nocturnal feeders. They cause a loss of appetite, a loss of body weight, and lowered production. Loss of blood may result in anemia. The ticks may also carry diseases. Heavy infestation may lead to so-called tick paralysis.

Malathion spray applied at the rate of 1 gallon per 1,000 square feet of building surface is the recommended treatment. Use according to the Manufacturer's recommendations.

the bed bug

The bed bug is a brown or yellow-red bug found under roosts and in cracks and crevices. The bugs are nocturnal feeders. They do not remain on the birds during the daylight hours. The birds pull out their own feathers in an effort to relieve the intense itching caused by the bed bugs. The treatment for this parasite is the same as for the northern fowl mite.

scaly leg mites

This mite is microscopic in size. It burrows into the skin of the legs, under the scales and may occasionally attack the comb and wattles. It may cause the birds to go lame or be crippled for life. There is evidence of scales and crusts on the legs, and the legs may be swollen. Scaly leg mite is not common where proper sanitation is practiced.

Treatment is on an individual bird basis. Dip legs in a mixture of 1 part kerosene plus 2 parts raw linseed oil.

chiggers

The chigger that attacks poultry is the same red mite that attacks man. Birds that are infested with chiggers may become droopy and emaciated. Chiggers attach themselves to the skin of the bird in clusters under the wings and on the back and neck. They may cause abcesses and inflamed areas. Birds may even die with severe infestations.

One aid in controlling chiggers is to keep the grass and weeds on range or in the yards cut short. The range may also be dusted with 1% malathion at the rate of 40 pounds per acre.

precautions for drug and pesticide use

Drugs and pesticides should be used with caution. To get optimum production results from poultry, the use of a drug or pesticide is sometimes indicated. However, before they are used make certain of the problem and then use the material of choice according to recommendations. Never use a material that is not registered for use on poultry.

Neither drugs nor pesticides are intended as substitutes for good preventive management. They work best when used in combination with good sanitation and sound management practices. Early diagnosis and treatment of disease or parasite problems is to be recommended.

Agencies of the federal government have set up definite withdrawal periods and tolerance levels for various materials used in poultry. The Food and Drug Administration, for example, has set minimum periods of time certain drugs must be withdrawn prior to poultry slaughter. These withdrawal periods vary from 1 or 2 days up to several days and are subject to change from time to time. Maximum tolerances for residues of certain materials are also established. Some insecticides can be used around poultry but not directly on the birds or on the eggs or in the nests. Some pesticide materials cannot be used within a certain number of days prior to slaughter. Frequency of use is also an important consideration with some materials.

Withdrawal periods and tolerances, as mentioned above, do change, and accepted treatment materials change, so the specific precaution for use of various materials will not be dealt with here. However, it is hoped that the point has been strongly made that drugs and insecticides should be used discriminately, following all the precautions on the label for the use of a given material. Improperly used, they can be injurious to man, animals and plants.

Store all drugs and pesticides in the original containers in a locked storage area. Keep them out of the reach of children and animals. Avoid the prolonged inhalation of sprays or dusts, and wear protective clothing and equipment where recommended. Be Safe!

Nutritional Diseases

A number of diseases of poultry may be caused by nutritional deficiencies or imbalances. With the well formulated diets on the market today, nutritional problems are very infrequent. Therefore, a thorough discussion of nutritional diseases will not be undertaken. However, to point out the importance of good nutrition, it may be well to just mention a few of the more common nutritional diseases.

rickets

Rickets is caused by a deficiency of Vitamin D3. Birds with rickets appear to be weak, have stiff swollen joints, soft beaks, soft bones of the leg and enlarged rib bones. A deficiency of phosphorous can also cause rickets.

perosis

Perosis is a leg weakness usually caused by a deficiency of manganese in the diet. Choline, niacin and biotin may also have a role. With this deficiency, most chicks will respond to manganese fortification in the feed.

encephalomalacia

Encephalomalacia or Vitamin E deficiency is commonly called "crazy chick disease" because it causes staggering, uncoordination, and paralysis of affected chicks. Vitamin E deficiency causes a part of the brain to degenerate. Positive diagnosis of this disease requires a study of the brain tissues.

curly toe paralysis

Curly Toe Paralysis is caused by a riboflavin deficiency in the feed. The birds become unthrifty and uneconomical. One or both feet may be affected. Additional fortification of the feed with riboflavin will help to prevent further development of this disorder.

nutritional roup — vitamin A deficiency

Birds that are deficient in Vitamin A become very unthrifty. Exudates may form in the eyes and nostrils and whittish pustules develop in the mouth, esophagus and crop. Mortality in young poultry can be very high if the deficiency is not corrected. Adult birds exhibit reduced production, lowered hatchability and decreased livability in the offspring. Yellow corn, legumes, grasses and fish oils are good sources of Vitamin A.

Predators

rats and mice

Rats and mice are a common problem on poultry farms. They like poultry feed, eat it, and destroy it. They may destroy feed bags if they are not hung up where they cannot get at them. They destroy insulation in poultry buildings and are carriers of disease. A few rats or mice will quickly produce a large population if not controlled.

Rats and mice like hiding places. Given places to hide in or around the chicken house, they will set up housekeeping. They like to inhabit the space between double sheathed walls, stone foundations, under floors and in and around junk. Good housekeeping will help to eliminate hiding spots and prevent rodent problems.

Trapping can be helpful where the populations of rats and mice are small. Keeping a cat in the area will also help to control them. However, baiting is the best way of handling large populations.

The anti-coagulent baits are the most effective means of erradicating rats and mice. Anti-coagulents cause the rodents to bleed to death internally after they have consumed the bait for several days.

To be effective, bait stations *(Figure 36)* need to be set up near walls, burrows and paths that the rats and mice tend to follow. These bait stations need to be kept full of fresh bait — but no more than they will eat in 2 or 3 days. If bait gets old and musty, they will not touch it. Fresh water should be located near the baiting stations to increase consumption of the bait.

Ideally, grain should be stored in rat and mouse proof bins or containers. Covered galvanized garbage cans are excellent for storing feed for small flocks.

other predators

Although rats and mice are the biggest nuisance to small flocks, one should be aware of other possible problems.

At times, dogs and cats may raid the poultry yard or house. Cats can be a problem if they gain access to a pen of young chicks. Occasionally dogs get into the poultry house or poultry yard and manage to kill young or adult birds. Usually, the real loss comes from piling of the birds and subsequent suffocation due to fright.

Foxes and raccoons frequent areas where chickens are kept. The work

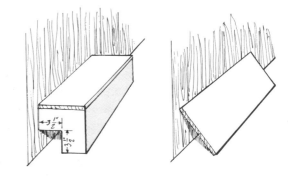

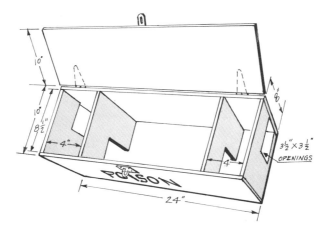

(Figure 36. Various types of bait stations.)

186

of a fox can frequently be identified by the signs of hair on the fence, or a hole where he crawled through. He usually kills several birds, and frequently leaves them behind partially buried. When the raccoon comes to visit he tends to eat the crops out of the birds, and some heads may be missing. Given the opportunity, he returns every fourth or fifth night to get more.

Owls also like chicken. Where one or two birds are killed every night and the heads and necks are missing, an owl may be the problem.

When birds are killed occasionally and are badly beaten, a dog problem is indicated. If several birds are killed on an irregular basis, it may also be the work of a mink or weasel, in which case there will be small bites around the neck and head.

Confinement-rearing has prevented many of the losses from predators. When birds are ranged or yarded they should be closed in a shelter at night. If range shelters are used, the space beneath the shelter should be screened. The installation of an electric fence a few inches above the ground may serve to discourage some predators.

Use steel jump traps on top of fence posts or high poles to catch owls and hawks. They like to land on high objects and select their prey. Sometimes the most effective means of protecting the poultry flock is to get rid of the predator causing the problem. It is always well to check local laws governing shooting and trapping wild animals, birds, or your neighbor's dog or cat.

List of Diagnostic Laboratories
by States

ALABAMA

State of Alabama Veterinary Diagnostic Laboratory
P.O. Box 127
Albertville, Alabama 35950

State of Alabama Veterinary Diagnostic Laboratory
P.O. Box 2209
Auburn, Alabama 36830

State of Alabama Veterinary Diagnostic Laboratory
192 Boswell Street
Elba, Alabama 36323

ALASKA

Alaska State Federal Laboratory
P.O. Box 807
Palmer, Alaska 99645

ARIZONA

Department of Animal Pathology
University of Arizona
Tucson, Arizona 95721

ARKANSAS

Arkansas Livestock and Poultry Commission Diagnostic Laboratory
2915 South Pine
Little Rock, Arkansas 72204

Arkansas Livestock and Poultry Commission Diagnostic Laboratory
Highway 71 N. Box 766
Springdale, Arkansas 72764

CALIFORNIA

California Department of Agr. Veterinary Lab. Services
2789 South Orange Avenue
Fresno, California 93725

California Department of Agr. Veterinary Lab. Services
1500 Petaluma Blvd. S.
Petaluma, California 94952

California Department of Agr. Veterinary Lab. Services
3290 Meadowview Road
Sacramento, California 95823

San Diego County Veterinary Laboratory
5555 Overland Avenue
San Diego, California 92123

California Department of Agr. Veterinary Lab. Services
P.O. Box 255
San Gabriel, California 91778

Poultry Pathological Laboratory
P.O. Box P
Turlock, California 95380

COLORADO

Colorado State University, Animal Disease Diagnostic Laboratory
Ark. Valley Br. Sta.,
Rocky Ford, Colorado 81067

CONNECTICUT

Department of Animal Diseases
Box U-89
University of Connecticut
Storrs, Connecticut 06268

DELAWARE

Avian Disease Laboratory
Route 2, Box 47,
Georgetown, Delaware 19947

Eastern Shore Laboratory
P.O. Box 657
Laurel, Delaware 19956

Sterwin Laboratories Incorporated
Millsboro, Delaware 19966

Dept. of Animal Science and Agricultural Biochemistry
University of Delaware
South College Avenue
Newark, Delaware 19711

FLORIDA

Dade City Diagnostic Laboratory
P.O. Box 1031
Dade City, Florida 33525

Kissimmee Animal Disease Diagnostic Laboratory
P.O. Box 460
Kissimmee, Florida 32741

Live Oak Diagnostic Laboratory
P.O. Drawer O,
Live Oak, Florida 32060

Miami Springs Diagnostic Laboratory
8701 N.W. 58th Street
Miami, Florida 33166

GEORGIA

Veterinary Pathology
Georgia Department of Agriculture
19 Hunter Street SW
Atlanta, Georgia 30334

Georgia Poultry Laboratory Number 2
P.O. Box 349
Canton, Georgia 30114

Georgia Poultry Laboratory Number 4
1126 Lamar Street
Dalton, Georgia 30720

Georgia Poultry Laboratory Number 1
P.O. Box 148
Oakwood, Georgia 30566

North East Georgia Poultry Laboratory
Box 37
Royston, Georgia 30662

Georgia Poultry Laboratory Number 6
Coastal Pl. Expt. Sta.
Tifton, Georgia 31794

HAWAII

Hawaii Dept. of Agr., Veterinary Laboratory Branch
1428 South King Street
Honolulu, Hawaii 96814

IDAHO

Idaho Livestock Disease Control Laboratory
P.O. Box 7249
Boise, Idaho 83707

ILLINOIS

Regional Diagnostic Laboratory
235 North Walnut Street
Centralia, Illinois 62801

Peoria Regional Diagnostic Laboratory
P.O. Box 1078
Peoria, Illinois 61601

Laboratories of Veterinary Diagnostic Medicine
University of Ill.
Urbana, Illinois 61801

IOWA

Veterinary Diagnostic Laboratory
Iowa State University
Ames, Iowa 50010

Dr. Mayfield Laboratories Incorporated
1209 South Main St.
Charles City, Iowa 50616

Salsbury Laboratory (Veterinary Services)
2000 Rockford Road
Charles City, Iowa 50616

Diagnostic Laboratory, Fort Dodge Laboratory Inc.
800 5th Street N.S.
Fort Dodge, Iowa 50501

KANSAS

Veterinary Diagnostic Laboratory
College of Vet. Med.
Manhattan, Kansas 66502

KENTUCKY

Kentucky Dept. of Agr., Animal Diagnostic Laboratory
North Drive
Hopkinsville, Kentucky 42240

Dept. of Veterinary Science Diagnostic Laboratory
University of Kentucky
Lexington, Kentucky 40506

Central Kentucky Animal Disease Laboratory
RR 6 Newtown Pike
Lexington, Kentucky 40505

LOUISIANA

W.E. Anderson Livestock Diagnostic Laboratory
P.O. Box 1276 Plank Road
Baton Rouge, Louisiana 70821

Central Louisiana Livestock Diagnostic Laboratory
Route 1 Box 51-F
Lecompte, Louisiana 71346

Northwest Louisiana Livestock Diagnostic & Research Laboratory
Robeline Rd. Box 4297 N.S.U.
Natchitoches, Louisiana 71457

MAINE

University of Maine Animal Pathology Diagnostic & Research Laboratory
Hitchner Hall
Orono, Maine 04473

Maine Poultry Consultants
Box 262
Waterville, Maine 04901

MARYLAND

  Animal Health Department Laboratory
  RFD
  Centreville, Maryland 21617

  Animal Health Department Laboratory
  P.O. Box 1234
  Frederick, Maryland 21701

  Animal Health Department Laboratory
  12 South Third St.
  Oakland, Maryland 21550

  Animal Health Department Laboratory
  P.O. Box J.
  Salisbury, Maryland 21801

MASSACHUSETTS

  Dept. of Veterinary & Animal Sciences
  Paige Laboratory
  University of Mass.
  Amherst, Mass. 01002

  Avian Diagnostic Laboratory
  240 Beaver Street
  Waltham, Mass. 02154

MICHIGAN

  Veterinary Diagnostic Laboratory
  Michigan State University
  East Lansing, Michigan 48823

  Michigan Dept. of Agr. Laboratory Division
  1615 S. Harrison Road
  East Lansing, Michigan 48823

  Escanaba Laboratory, Michigan Dept. of Agriculture
  305 Ludington Street
  Escanaba, Michigan 49829

MINNESOTA

  University of Minnesota, Vet. Diagnostic Laboratory
  St. Paul Campus,
  St. Paul, Minnesota 55101

MISSISSIPPI

Veterinary Diagnostic Laboratory,
Mississippi Board of Animal Health
2531 North West St.
Jackson, Mississippi 39201

MISSOURI

Veterinary Medical Diagnostic Laboratory
University of Missouri
Columbia, Missouri 65201

Ralston Purina Veterinary Laboratory
Veterinary Department
Checkerboard Square
St. Louis, Missouri 63188

MONTANA

State of Montana, Livestock Sanitary Board Diagnostic Laboratory
P.O. Box 997
Bozeman, Montana 59715

NEBRASKA

Harris Laboratories Incorporated
P.O. Box 80837
Lincoln, Nebraska 68502

Diagnostic Laboratory, Department of Veterinary Science
University of Nebraska
Lincoln, Nebraska 68503

Veterinary Science Laboratory
University of Nebraska
North Platte Station
North Platte, Nebraska 69101

NEVADA

Animal Disease Laboratory
Nevada Department of Agriculture
350 Capitol Hill Ave.
Reno, Nevada 89502

NEW HAMPSHIRE

Veterinary Diagnostic Laboratory
University of New Hampshire
Durham, New Hampshire 03824

NEW JERSEY

Rutgers Poultry Diagnostic Laboratory
Thompson Hall
New Brunswick, New Jersey 08903

New Jersey Animal Health Diagnostic Laboratory
John Fitch Plaza
Trenton, New Jersey 08625

Rutgers Poultry Health Laboratory
2569 Landis Avenue
Vineland, New Jersey 08360

NEW MEXICO

State Federal Cooperative Laboratory
P.O. Box 464
Albuquerque, New Mexico 87103

NEW YORK

Cornell University, Duck Research Laboratory
Box 217 Old Ctry. Road
Eastport L.I. New York 11941

Department of Avian Diseases
N.Y. State Vet. College
Ithaca, New York 14850

Regional Poultry Disease Laboratory
88 Prince Street
Kingston, New York 12401

NORTH CAROLINA

Livestock and Poultry Disease Diagnostic Laboratory
Box 5658 Western Blvd.
Raleigh, North Carolina 27606

NORTH DAKOTA

North Dakota State Veterinary Diagnostic Laboratory
North Dakota State University
Fargo, North Dakota 58102

OHIO

Ohio State University, Dept. Veterinary Clinical Sciences, Clinical Pathology
Laboratory
2578 Kenny Road
Columbus, Ohio 43210

Ohio Department of Agriculture, Animal Disease Diagnostic Laboratory
Reynoldsburg, Ohio 43068

OKLAHOMA

State Federal Laboratory
50 Northeast 23 Street
Oklahoma City, Oklahoma 73105

College of Veterinary Medicine
Oklahoma St. University
Stillwater, Oklahoma 74074

OREGON

Oregon State Veterinary Diagnostic Laboratory
Oregon State University
Corvallis, Oregon 97331

State Federal Veterinary Laboratory
635 Capitol St. N.E.
Salem, Oregon 97310

PENNSYLVANIA

Regional Poultry Diagnostic Laboratory
Pennsylvania Department of Agriculture
Doylestown, Pennsylvania 18901

Southwest District Diagnostic Laboratory
R. 1. Evans City, Pennsylvania 16033

Poultry Diagnostic Laboratory
New Bolton Center
Kennett Square, Pennsylvania 19348

Whitmoyer Laboratories Incorporated
19 N. Railroad Street
Myerstown, Pennsylvania 17067

Pennsylvania Dept. of Agriculture, Bureau of Animal Industry
Laboratory
Summerdale, Pennsylvania 17093

Regional Diagnostic Laboratory
R 92 South
Tunkhannock, Pennsylvania 18657

Poultry Diagnostic Laboratory, Pennsylvania State University
Wiley Laboratory
University Park, Pennsylvania 16802

Animal Disease Diagnostic Laboratory
Animal Disease Building
University Park, Pennsylvania 16802

PUERTO RICO

Puerto Rico Animal Diagnostic Laboratory
P.O. Box E
Dorado, Puerto Rico 00646

RHODE ISLAND

Diagnostic Laboratory Dept. of Animal Pathology
University of Rhode Island
Kingston, Rhode Island 02881

SOUTH CAROLINA

Clemson University, Livestock Laboratory
P.O. Box 1771
Columbia, South Carolina 29201

SOUTH DAKOTA

Animal Disease Research and Diagnostic Laboratory
South Dakota State University
Brookings, South Dakota 57006

TENNESSEE

Poultry Service Laboratory
Knoxville, Tennessee 37901

C.E. Kord Animal Disease Laboratory
P.O. Box 40627 Mel. Sta.
Nashville, Tennessee 37204

TEXAS

Texas A and M University, Poultry Disease Laboratory
FM 108 Box 84
Gonzales, Texas 78629

UTAH

Utah State University Veterinary Diagnostic Laboratory
Utah State University
Logan, Utah 84321

Branch Veterinary Laboratory
Utah Agricultural Experiment Station
P.O. Box 1068
Provo, Utah 84601

State Chemist Office and State Federal Cooperative Laboratory
412 Capitol Building
Salt Lake City, Utah 84114

VERMONT

Animal Pathology Laboratory
Hills Science Building
University of Vermont
Burlington, Vermont 05401

VIRGINIA

Department of Veterinary Science
Virginia Polytechnic and State University
Va. Polytech. Inst.
Blacksburg, Virgina 22801

Division of Animal Health and Dairies Regulatory Laboratory
116 Reservoir Street
Harrisonburg, Virginia 22801

Division of Animal Health and Dairies Regulatory Laboratory
Ivor, Virginia 23866

Division of Animal Health and Dairies Regulatory Laboratory
Box 4191
Lynchburg, Virginia 24502

Division of Animal Health and Dairies Central Laboratory
1 North 14 Street
Richmond, Virginia 23219

Division of Animal Health and Dairies Regulatory Laboratory
234 West Shirley Avenue
Warrenton, Virginia 22186

Division of Animal Health and Dairies Regulatory Laboratory
Box 436
Wytheville, Virginia 24382

WASHINGTON

Department of Veterinary Pathology
Pullman, Washington 99163

Poultry Diagnostic Laboratory
Western Washington Research & Experiment Center
Puyallup, Washington 98371

WEST VIRGINIA

State Federal Cooperative Animal Health Laboratory
RM B 86 Capitol Bldg.
Charleston, West Virginia 25305

West Virginia
Regional Animal Health Laboratory
Box 568
Moorefield, West Virginia 26836

West Virginia University, Division of Animal & Veterinary Sciences
Agricultural Science Building EV. Camp.
Morgantown, West Virginia 26506

WISCONSIN

Regional Animal Health Laboratory
1418 LaSalle Avenue
Barron, Wisconsin 54812

Murphy Diagnostic Laboratory
124 South Dodge Street
Burlington, Wisconsin 53105

Central Animal Health Laboratory
6101 Mineral Pt. Rd.
Madison, Wisconsin 53705

WYOMING

Wyoming State Veterinary Laboratory
Box 950
Laramie, Wyoming 82070

Addresses of Extension Poultry Specialists
by States

ALABAMA

Auburn University, Auburn, Alabama 36830
102 Federal Bldg. P.O. Box 15, Cullman, Alabama 35055

ARIZONA

Agricultural Science Building, Univ. of Arizona, Tucson, Arizona 85721

ARKANSAS

University of Arkansas, P.O. Box 391, Little Rock, Arkansas 72203
University of Arkansas, Dept. Animal Sci. R-C123, Fayetteville, Arkansas
72701

CALIFORNIA

University of California, Department of Avian Sciences, Davis, California
95616
University of California, 306 Agricultural Extension Building, Riverside,
California 92502
P.O. Box 1411, Modesto, California 95353
Agr. Res. & Ext. Center, 9240 S. Riverbend Ave., Parlier, California 93648
Suite 202, 21160 Box Springs Rd., Riverside, California 92507
566 Lugo Avenue, San Bernardino, California 92415
Building 4, 5555 Overland Avenue, San Diego, California 92123
Room 100-P, Mendocino Avenue, Santa Rosa, California 95401
684 Buena Vista St., Ventura, California 93001
1000 S. Harbor Blvd., Anaheim, California 92805

COLORADO

Colorado State University, Fort Collins, Colorado 80521

CONNECTICUT

University of Connecticut, Storrs, Connecticut 06268
Agricultural Center, Haddam, Connecticut 06492
24 Hyde Avenue Rt. 30, Rockville, Connecticut 06066
562 New London Turnpike, Norwich, Connecticut 06360

DELAWARE
    University of Delaware, RD 2, Box 48, Georgetown, Delaware 19947
    Agr. Hall, Newark, Delaware 19711

FLORIDA

    University of Florida, Gainesville, Florida 32603
    Chipley, Florida 32428

GEORGIA

    University of Georgia, Athens, Georgia 30602
    Calhoun, Georgia 30701
    Oakwood, Georgia 30566
    Tifton, Georgia 31794

HAWAII

    University of Hawaii, 1825 Edmondson Rd., Honolulu, Hawaii 96822

IDAHO

    University of Idaho, Moscow, Idaho 83843

ILLINOIS

    University of Illinois, 322 Mumford Hall, Urbana, Illinois 61801

INDIANA

    Purdue University, Lafayette, Indiana 47907

IOWA

    Iowa State University, Kildee Hall, Ames, Iowa 50010

KANSAS

    Kansas State University, Leland Call Hall, Manhattan, Kansas 66506

KENTUCKY

    University of Kentucky, Lexington, Kentucky 40506
    Somerset Community Col., 808 Monticello Rd., Somerset, Kentucky 42501
    1270 Montgomery Avenue, Ashland, Kentucky 41101

LOUISIANA

Louisiana State University, Knapp Hall, Baton Rouge, Louisiana 70803

MAINE

University of Maine, Hitchner Hall, Orono, Maine 04473
P.O. Building, Belfast, Maine 04915
P.O. Building, Lewiston, Maine 04241
Federal Bldg. Rm. 209, Rockland, Maine 04841

MARYLAND

University of Maryland, Dept. of Poultry Science, College Park, Maryland 20742
Broiler Sub-Station, RFD 5, Salisbury, Maryland 21801

MASSACHUSETTS

University of Mass., 314 Stockbridge Hall, Amherst, Massachusetts 01002

MICHIGAN

Michigan State University, 113 Anthony Hall, East Lansing, Michigan 48823
Coop Extension Service, P.O. Box 79, Zeeland, Michigan 49464

MINNESOTA

University of Minnesota, St. Paul, Minnesota 55101

MISSISSIPPI

Mississippi State University, P.O. Box 5425, Mississippi State, Mississippi 39762
Box 9714, Jackson, Mississippi 39206

MISSOURI

University of Missouri, Columbia, Missouri 65201
RDI Bldg., Belcrest & E. Trafficway, Springfield, Missouri 65802

NEBRASKA

University of Nebraska, Lincoln, Nebraska 68503

NEW HAMPSHIRE

University of New Hampshire, 55 Pleasant St., Rm. 331, Concord, N.H. 03301
Kendall Hall, Durham, New Hampshire 03824

NEW JERSEY

Rutgers — The State University, CAES P.O. Box 231, New Brunswick, New Jersey 08903

NEW MEXICO

New Mexico State University, Dept. Poultry Sci., Box 3P, Las Cruces, New Mexico 88001

NEW YORK

Cornell University, Rice Hall, Ithaca, New York 14850
249 Highland Ave., Rochester, New York 14620
380 Federal Building, Syracuse, New York 13202
Coop Ext. Regional Office, Martin Rd., Voorheesville, N.Y. 12186

NORTH CAROLINA

North Carolina State University, P.O. Box 5307 Scott Hall, Raleigh, N.C. 27607

NORTH DAKOTA

North Dakota State Univ. Stevens Hall, Rm. 218A, Fargo, North Dakota 58102

OHIO

Ohio State Univ., 2120 Fyffe Road, Columbus, Ohio 43210
2120 Fyffe Road, Columbus, Ohio 43210 (Ohio State Univ.

OKLAHOMA

Oklahoma State University, Stillwater, Oklahoma 74074

OREGON

Oregon State University, Poultry Sci. Dept., Corvallis, Oregon 97331

PENNSYLVANIA

Pennsylvania State University, University Park, Pennsylvania 16802

PUERTO RICO

University of Puerto Rico, Rio Piedras, Puerto Rico 00928

RHODE ISLAND

University of Rhode Island, Kingston, Rhode Island 02881

SOUTH CAROLINA

Clemson University, Clemson, South Carolina 29631
P.O. Box 378, York, South Carolina 29745
P.O. Box 1711, Columbia, South Carolina 29201

SOUTH DAKOTA

South Dakota State University, Brookings, South Dakota 57006

TENNESSEE

University of Tennessee, P.O. Box 1071, Knoxville, Tennessee 37901

TEXAS

Texas A & M University, College Station, Texas 77843
Res. & Ext. Center, Box 220, Overton, Texas 75684

UTAH

Utah State University, Logan, Utah 84322

VERMONT

University of Vermont, Burlington, Vermont 05401

VIRGINIA

Virginia Polytechnic Institute, Blacksburg, Virginia 24061

WASHINGTON

Washington State University, Pullman, Washington 99163
Western Washington, Res. & Ext. Center, Puyallup, Washington 98371

WEST VIRGINIA

West Virginia University, Morgantown, West Virginia 26506

WISCONSIN

University of Wisconsin, Dept. Poultry Sci., Col. of Agr. & Life Sci.,
1675 Observatory Dr., Madison, Wisconsin 53706

WYOMING

University of Wyoming, Univ. Station, P.O. Box 3354, Laramie, Wyoming
82071

WASHINGTON, D.C.

Extension Service — USDA, Room 5509 South Building, Washington, D.C.
20250

Directory of Poultry Publications

A number of U.S.D.A. and State Extension publications on poultry production and marketing are available and may be obtained by contacting the local County Extension Agent and State or Area Poultry Extension Specialists. The addresses of the Poultry specialists, by states, are given in a preceding section. A partial list of trade publications and books follows:

TRADE PUBLICATIONS

Broiler Industry, Garden State Building, Sea Isle City, N.J. 08243 — Monthly

Egg Industry, Garden State Building, Sea Isle City, N.J. 08243 — Monthly

Feedstuffs, 2501 Wayzata Blvd., Minneapolis, Minn. 55404 — Weekly

Poultry Digest, Garden State Building, Sea Isle City, N.J. — Monthly

Poultry Meat, Watt Publishing Co., Mount Morris, Illinois 61054 — Monthly

Poultry Tribeun, Watt Publishing Co., Mount Morris, Illinois 61054 — Monthly

The Poultry Times, P.O. Box 1338, Gainsville, Georgia 30501 — Weekly

Turkey World, Watt Publishing Co., Mount Morris, Illinois 61054 — Monthly

BOOKS

Commercial Chicken Production Manual, Mack O. North, 1972. The AVI Publishing Co. Inc.
Diseases of Poultry, Biester & Schwartz, 5th Edition, 1965 Iowa State University

Farm Poultry Production, Wilson and Card, Interstate

Marketing Poultry Products, Benjamin, Gwin, Faber and Termohlen, 5th Ed. 1960 Wiley

Modern Waterfowl Management and Breeding Guide, Oscar Grow 1972, American Bantam Assoc.

Nutrition of the Chicken, Scott, Nesheim & Young, 1969 M. L. Scott and Associates

Poultry Production, Leslie E. Card, 10th Edition, 1966, Leo & Febiger.

Standard of Perfection, American Poultry Association, 1968, Cooperative Publishing Co.

The Avian Egg, Romanoff, 1st Edition, 1949, Wiley

The Avian Embryo, Romanoff, 1959, MacMillan

Turkey Management, Marsden and Martin, 6th Edition 1955, Interstate

PUBLISHERS' ADDRESSES

American Bantam Association, P.O. Box 464, Chicago, Ill. 60690

The AVI Publishing Company, Inc., Westport, Connecticut 06880

Cooperative Publishing Co., Guthrie, Oklahoma 73044

The Interstate Printers & Publishers, Inc., Danville, Ill. 61832

Iowa State University Press, Press Bldg., Ames Iowa 50010

Lea and Febiger, 600 Washington Square, Philadelphia, Pa. 19105

The Macmillan Co., Riverside, N.J. 08075

M. L. Scott and Associates, Ithaca, N.Y. 14850

John Wiley and Sons, Inc. Publishers, 605 Third Ave., New York, N.Y. 10016

Sources of Supplies and Equipment

Many of the supplies and equipment needs for the small poultry flock can be found at local feed store, hatchery or other agricultural supply outlets. Some of the large mail order houses such as Sear's and Ward's also handle agricultural supplies

As an aid to locating sources of certain items not readily available in some areas, a partial list of supplies and equipment and possible sources is included below.

Space would not permit a complete listing of products nor company names. There is no intent to recommend one source over another.

Bags — Dressed Poultry
Dow Chemical Co. Flexible Packaging Sales, James Savage Building., Midland, Michigan 48640

W. R. Grace and Co., Cryovac Division, Box 464, Duncan, South Carolina 29334

Gordon Johnson Industries, 2519 Madison Ave., Kansas City, Missouri 64108

Mobil Chemical Co., Plastics Div. Tech Center, Macedon, New York 14502

Bands — Identificiation — Leg and Wing
Gey Band and Tag Co. Inc., P.O. Box 363, Norristown, Pennsylvania 19401

National Band & Tag Co., 721 York Street, Newport, Kentucky 41071

Brooders
Anderson Box Co., P.O. Box 1851, Indianapolis, Indiana 46206

Bramco Products, Mariette St., Canton, Georgia 30114

Brower Manufacturing Co., Box 1123, Quincy, Illinois 62301

Lyon Electric Co., 3425 Hancock St., P.O. Box 81303, San Diego, California 92138

Shenandoah Mfg. Co. Inc., Edom Rd., P.O. Box 839, Harrisonburg, Virginia 22801

Big Dutchman, A div. of U.S. Industries Inc., 200 N. Franklin St., Zeeland, Michigan 49464

Cages
American-Can, Div. of Jamesway Co. Lts., P.O. Box 156, Bowling Green, Ohio 43402

Anderson Box Co., P.O. Box 1851, Indianapolis, Indiana 46206

Big Dutchman, A div. of U.S. Industries Inc., 200 N. Franklin St., Zeeland, Michigan 49464

Diamond International Corp., 23400 Haggerty Road, Farmington, Michigan 48024

Chore-Time Equipment Inc., P.O. Box 518, Milford, Indiana 46542

Cannibalism Supplies and Equipment — Specs
Gey Band and Tag Co. Inc., P.O. Box 363, Norristown, Pennsylvania 19401

Kuhl Corporation, Box 26, Flemington, New Jersey 08820

National Band & Tag Co., 721 York Street, Newport, Kentucky 41071

Cannibalism Supplies and Equipment — Debeakers
Lyon Electric Co., 3425 Hancock St., P.O. Box 81303, San Diego, California 92138

Cannibalism Supplies and Equipment — Ointments and Paints
Eastern Laboratories Inc., 1483 Washington Ave., Vineland, New Jersey 08360
Vineland Laboratories Inc., 2284 E. Landis Ave., Vineland, New Jersey 08360
Caponizing Instruments
Mattox & Moore Inc., 1503 E. Riverside Dr., Indianapolis, Indiana 46202
Thomas Equipment & Supply Co., 5505 Langley Rd., Houston, Texas 77016
Catching Hooks
Holland Co., Inc., 208 Market St., Box 645, Lexington, North Carolina 27292
Thomas Equipment & Supply Co., 5505 Langley Rd., Houston, Texas 77016
Disinfectants
Carbola Chemical Co. Inc., Natural Bridge, New York 13665
The Diversey Corp., 212 E. Monroe St., Chicago, Illinois 60606
The Dow Chemical Co. Agri. Products, Box 1706, Midland, Michigan 48640
Hess & Clark, Div. of Rhodia Inc., 7th & Orange, Ashland, Ohio 44805
Pfizer Inc., Agri. Div., 235 E 42nd St., New York, N.Y. 10017
Salsbury Laboratories, Charles City, Iowa 50616
Vineland Laboratories Inc., 2285 E. Landis Ave., Vineland, New Jersey 08360
Whitmoyer Laboratories Inc., Sub. of Rohn & Haas Co., Myerstown, Pennsylvania 17067

Egg Baskets — Gathering
Barker Equipment Co., Div. Conco Inc., P.O. Box 111, Mendota, Illinois 61342
Big Dutchman, A div. of U.S. Industries Inc., 200 N. Franklin St., Zeeland, Michigan 49464
Brower Mfg. Co., Box 1123, Quincy, Illinois 62301
Kuhl Corporation, Box 26, Flemington, New Jersey 08820
Egg Candlers — Hand
Big Dutchman, A div. of U.S. Industries Inc., 200 N. Franklin St., Zeeland, Michigan 49464
Brower Mfg. Co., Box 1123, Quincy, Illinois 62301
Egg Cartons
Anderson Box Co., P.O. Box 1851, Indianapolis, Indiana 46206
Continental Packaging Corp., 555 North Michigan Ave., Kenilworth, New Jersey 07033
Diamond International Corp., Diamond Fiber Products Div., 733 Third Ave., New York, N.Y. 10017
Keyes Fibre Company, 160 Summit Ave., Montvale, New Jersey 07645
Packaging Corporation of America, 300 W. Main St., Griffith, Indiana 46319
Egg Cases and Filler Flats
Anderson Box Co., P.O. Box 1851, Indianapolis, Indiana 46206
Peter Berg & Company Inc., 917-19 Fifth St., N., Minneapolis, Minnesota 55401
Brooklyn Egg Case Co., Rt. 17 Box A., Harris, New York 12742
Diamond International Corp., Diamond Fiber Products Div., 733 Third Ave., New York, N.Y. 10017
Packaging Co. of California, 2900 San Pablo Ave., Berkeley, California 94702
Packaging Corporation of America, 300 W. Main St., Griffith, Indiana 46319

Egg Scales — Small
Brower Manufacturing Co., Box 1123, Quincy, Illinois 62301
Beacon Steel Products Co., Inc., Box 600 Westminster, Maryland 21157
W. Kent Co., 13145 Coronado Dr., North Miami, Florida 33161
National Poultry Equipment Co., Box 238, 615 Wells Ave., N. Renton, Washington 98055

Egg Washers — Small
Big Dutchman, A div. of U.S. Industries Inc., 200 N. Franklin St., Zeeland, Michigan 49464
National Poultry Equipment Co., Box 238, 615 Wells Ave., N. Renton, Washington 98055

Egg Washing Detergent — Sanitizers
Big Dutchman, A div. of U.S. Industries Inc., 200 N. Franklin St., Zeeland, Michigan 49464
Brower Manufacturing Co., Box 1123, Quincy, Illinois 62301
Kuhl Corporation, Box 26, Flemington, New Jersey 08820
National Poultry Equipment Co., Box 238, 615 Wells Ave., N. Renton, Washington 98055
Proctor & Gamble Agr. Chemicals, Sales Dept., Box 599, Cincinnati, Ohio 45201
Vineland Laboratories Inc., 2285 E. Landis Ave., Vineland, N.J. 08360

Fans — Ventilation
Aerovent Fan & Equipment Inc., 5530 Pennsylvania Ave., Box 9007, Lansing, Michigan 48909
Amer-Can, Div. of Jamesway Co. Ltd., P.O. Box 156, Bowling Green, Ohio 43402
Big Dutchman, A div. of U.S. Industries Inc., 200 N. Franklin St., Zeeland, Michigan 94964

Feeders & Waterers
Amer-Can, Div. of Jamesway Co. Ltd., P.O. Box 156, Bowling Green, Ohio 43402
Big Dutchman, A div. of U.S. Industries Inc., 200 N. Franklin St., Zeeland, Michigan 49464
Bramco Products, Marietta St., Canton, Georgia 30114
Brower Manufacturing Co., Box 1123, Quincy, Illinois 62301
H. D. Hudson Mfg. Co., 154 E. Erie St., Chicago, Illinois 60611
Shenandoah Mfg. Co. Inc., Edom Rd., P.O. Box 839, Harrisonburg, Virginia 22801

Incinerators
Big Dutchman, A div. of U.S. Industries Inc., 200 N. Franklin St., Zeeland, Michigan 49464
Diamond International Corp., Diamond Fiber Products Div., 733 Third Ave., New York, N.Y. 10017
Shenandoah Mfg. Co. Inc., Edom Rd., P.O. Box 839, Harrisonburg, Virginia 22801

Incubators — Small
Brower Manufacturing Co., Box 1123, Quincy, Illinois 62301
Jamesway, Div. of Butler Mfg. Co., 104 W. Milwaukee Ave., Ft. Atkinson, Wisconsin 53538
Humidaire Incubator Co., 217 W. Wayne St., New Madison, Ohio 45346
Lyon Electric Co., 3425 Hancock St., P.O. Box 81303, San Diego, California 92138

Petersime Incubator Co., Gettysburg, Ohio 45328

Robbins Incubator Co., Box 899, Denver, Colorado 80201

Knives — Poultry Dressing

Barker Poultry Equipment Co., P.O. Box 368, 802 South Madison Ave., Ottumwa, Iowa 52501

W. Kent Co., 13145 Coronado Dr., North Miami, Florida 33161

Thomas Equip. & Supply Co., 5505 Langley Rd., Houston, Texas 77016

Pickwick Company, P.O. Box 756, 1120 Glass Rd. N]E., Cedar Rapids, Iowa 52406

Nests

Amer-Can, Div. of Jamesway Co. Ltd., P.O. Box 156, Bowling Green, Ohio 43402

Anderson Box Co., P.O. Box 1851, Indianapolis, Indiana 46206

Brower Manufacturing Co., Box 1123, Quincy, Illinois 62301

Kuhl Corporation, Box 26, Flemington, New Jersey 08820

Shenandoah Mfg. Co. Inc., Edom Rd., P.O. Box 839, Harrisonburg, Virginia 22801

Poultry Pickers

Ashley Machine Inc., Box 2, Greensburg, Indiana 47240

Barker Poultry Equipment Co., P.O. Box 368, 802 South Madison Ave., Ottumwa, Iowa 52501

W. Kent Co., 13145 Coronado Dr., North Miami, Florida 33161

Pickwick Company, P.O. Box 756, 1120 Glass Rd. N.E., Cedar Rapids, Iowa 52406

Poultry Scalders

Ashley Machine Inc., Box 2, Greensburg, Indiana 47240

Barker Poultry Equipment Co., P.O. Box 368, 802 South Madison Ave., Ottumwa, Iowa 52501

W. Kent Co., 13145 Coronado Dr., North Miami, Florida 33161

Kuhl Corporation, Box 26, Flemington, New Jersey 08820

Pickwick Company, P.O. Box 756, 1120 Glass Rd. N.E., Cedar Rapids, Iowa 52406

Thomas Equipment & Supply Co., 5505 Langley Road, Houston Texas 77016

Thermostats — Mercury or Open Contact

Aerovent Fan & Equipment Inc., 5530 Pennsylvania Ave., Box 9007, Lansing, Michigan 48909

Big Dutchman, A div. of U.S. Industries Inc., 200 N. Franklin St., Zeeland, Michigan 49464

Chick Master Incubator Corp., 3212 W. 25th St., Cleaveland, Ohio 44109

Diamond International Corp., 23400 Haggerty Road, Farmington, Michigan 48024

Petersime Incubator Co., Gettysburg, Ohio 45328

Thermostats — Wafer

Brower Manufacturing Co., Box 1123, Quincy, Illinois 62301

Lyon Electric Co., 3425 Hancock St., P.O. Box 81303, San Diego, California 92138

Shenandoah Mfg. Co. Inc., Edom Rd., P.O. Box 839, Harrisonbufg, Virginia 22801

A. R. Wood-Northco, Box 218, Lucerne, Minnesota 56156

Time Clocks

Big Dutchman, A div. of U.S. Industries Inc., 200 N. Franklin St., Zeeland, Michigan 49464

Brower Manufacturing Co., Box 1123, Quincy, Illinois 62301

Diamond International Corp. Diamond Fiber Products Div., 733 Third Ave., New York, N.Y. 10017

Lyon Electric Co., 3425 Hancock St., P.O. Box 81303, San Diego, California 92138

Paragon Electric Co. Inc., 1600-12 St., Two Rivers, Wisconsin 54241

Water Warmers

Brower Manufacturing Co., Box 1123, Quincy, Illinois 62301

H. D. Hudson Mfg. Co., 154 E. Erie St., Chicago, Illinois 60611

Wax-Defeathering

Du Bois Chemicals, Du Bois Terrace, Cincinnati, Ohio 45202

Gainesville Machine Co. Inc., P.O. Box 1258, Gainesville, Georgia 30501

Gordon Johnson Industries, 2519 Madison Avenue, Kansas City, Missouri 64108

National Wax Co., 3650 Touhy, Skokie, Illinois 60076

REFERENCES

Aho, William A. and Talmadge, Daniel W., *Incubation and Embryology of the Chick,* Cooperative Extension Service, College of Agriculture and Natural Resources, University of Connecticut, Storrs, Conn.

Ash, William J., *Raising Ducks.* Farmers Bulletin, No. 2215, U.S.D.A. *Caponizing Chickens,* Leaflet 490, U.S.D.A.

Culling Hens, Farmers Bulletin No. 2216, Agricultural Research Service, U.S.D.A.

Duck Rations, Extension Stencil No. 25, Department of Poultry Science. Cornell University, Ithaca, N.Y.

Egg Grading Manual, Agricultural Handbook, No. 75, Consumer and Marketing Service, U.S.D.A.

Embryology and Biology of Chickens, The Vermont Extension Service, University of Vermont, Burlington, Vt.

Farm Poultry Management, Farmers Bulletin No. 2197, Agricultural Research Service, U.S.D.A.

Hauver, W. E. and Kilpatrick, L., *Poultry Grading Manual,* Agricultural Handbook No. 31, U.S.D.A.

Hunter, J. M. and Scholes, John C., *Profitable Duck Management,* The Beacon Milling Co., Cayuga, N.Y.

Jordan, H. C. and Schwartz, L. D., *Home Processing of Poultry,* Penn State University, College of Agr. Extension Service, University Park, Pa.

Marsden, Stanley J., *Turkey Production,* Agricultural Handbook No. 393, Agricultural Research Service, U.S.D.A.

Mercia, L. S., *Drawing Poultry,* 4-H Publication, Vermont Extension Service, University of Vermont, Burlington, Vt.

Mercia, L. S., *Killing, Picking and Cooling Poultry,* 4-H Publication, The Vermont Extension Service, University of Vermont, Burlington, Vt.

Mercia, L. S., *Lighting Replacements and Layers,* Vermont Extension Service, University of Vermont, Burlington, Vt.

Mercia, L. S., *The Family Laying Flock,* Brieflet 1218, Vermont Extension Service, University of Vermont, Burlington, Vt.

Moyer, D. D., *Virginia Turkey Management,* Publication 302, Extension Division, Virginia Polytechnic Institute, Blacksburg, Va.

Orr, H. L., *Duck and Goose Raising,* Publication 532, Ontario Department of Agr. and Food, Parliament Building, Toronto, Canada.

Ota, Hajime, *Housing and Equipment for Laying Hens,* Misc. Publication 728, Agr. Research Service, U.S.D.A.

1974-75 *Poultry Management and Business Analysis Manual,* Bulletin 566, Cooperative Extension Services, Extension Poultrymen in New England.

Raising Geese, Farmers Bulletin, No. 2251, Northeastern Region Agritural Research Service, U.S.D.A.

Raising Livestock on Small Farms, Farmers Bulletin No. 2224, U.S.D.A.

Ridlen, S. F., and Johnson, H.S., *4-H Poultry Management Manual — Unit 1,* Cooperative Extension Service, University of Illinois, Urbana, Ill.

Talmadge, Dan W., *Brooding and Rearing Baby Chicks,* Published by the Cooperative Extension Services of the Northeast.

Warren, Richard, *4-H Laying Flock Management,* 4-H Circular 79, Published by the Cooperative Extension Services in New England.

GENERAL FEED REQUIREMENTS
FOR DIFFERENT TYPES OF BIRDS
AT VARIOUS AGES

Type of Bird		Type Feed	Percent Protein	Lbs/100 Birds	
				White Egg	Brown Egg
Chickens:					
Pullets	0 — 8 wks.	Starter	20	280	364
	8 — 12 wks.	Grower	17	308	392
	12 — 22 wks.	Grower or Developer	14	1064	1289
Layers	22 + wks.	Layer	17	23	25
Broilers	0 — 4 wks.	Starter	22	200	
or	5 — 7 wks.	Grower	20	500	
Fryers	7 — 8 wks.	Finisher	18	100	
Roasters	0 — 4 wks.	Starter	22	200	
	4 — 13 wks.	Grower	16	1200	
	13 — market	Finisher	14	1000	
Turkeys:					
Small	0 — 7 wks.	Starter	28	600	
	7 — 18 wks.	Grower	20	3000	
Large	0 — 8 wks.	Starter	28	1000	
	8 — 16 wks.	Grower	20	3000	
	16 — 24 wks.	Finisher	14	3500	
Ducks: (White Pekin)					
For Market	0 — 2 wks.	Starter	22-24	214	
	2 — 4 wks.	Grower	18-20	510	
	4 — 8 wks.	Finisher	16-17	1410	
For Breeders					
	0 — 2 wks.	Starter	22-24	214	
	2 wks. maturity	Grower	18-20	*	
Geese: (White EMBDEN OR Chinese)					
	0 — 3 wks.	Starter	20-22	525	
	3 — 14 wks.	Grower	15	5600	
	14 wks. to market (24-30 wks.)	Grower	15		

Ducks and particularly geese are good foragers. From approximately 3 weeks of age with good forage available, savings of up to 35% in feed consumption can be realized.

A "NATURAL" LAYING HEN DIET*

To be fed as all-mash to medium size layers kept in floor pens

Ingredient	Amount (lbs.) 100 lbs.	Per Ton
Yellow corn meal	60.00	1200
Wheat middlings	15.00	300
Soybean meal (dehulled)	8.00	160
Maine herring meal (65%)	3.75	75
Meat & bone meal (47%)	1.00	20
Skim milk, dried	3.00	60
Alfalfa leaf meal (20%)	2.50	50
Iodized salt	0.40	8
Limestone, grd. (38% Ca)	6.35	127
Totals	100.00	2000

Calculated Analysis		Recommended (N.E.C.C.) (Per pound)
Metabolizable energy Cal./lb.	1252	1292
Protein	16.07	16
Lysine	0.79	0.74
Methionine	0.31	0.29
Methionine & cystine	0.55	0.54
Fat	3.67	3.33
Fiber	3.15	2.51
Calcium	2.77	2.75
Total phosphorus	0.53	0.50
Available phosphorus	0.44	0.42

Vitamins (units or mgs./lb.)

Vitamin A activity (U.S.P. units/lb.)	5112	5290
Vitamin D (I.C.U.)		1000
Riboflavin (mg.)	1.36	1.38
Pantothenic acid (mg.)	3.89	4.05
Choline Chloride (mg.)	411	500
Niacin (mg.)	17.46	16.95

The birds *must* receive *direct* sunlight to enable them to synthesize vitamin D. Unfortified cod liver oil can be fed in place of sunlight to supply vitamin D. The amount of cod liver oil would depend upon the potency of the oil — the need is for 1000 (I.C.U.) per pound of feed.

*(Prepared by Dr. Richard Gerry, Department of Animal & Veterinary Sciences, University of Maine, Orono, Maine)

POULTRY MANURE INFORMATION

Manure Production
Layers: 25 pounds per 100 per day with normal drying.
Four-tenths of a cubic foot per 100 per day.
Moisture content 75-80% as defecated.
Pollution load — 7 to 12 layers equal one human.
Weight of a cubic foot of poultry manure at 70% moisture is
approximately 65 pounds.

Broilers: Manure and litter per 100 birds
8 weeks equals 400 pounds or 11.2 cubic feet at 25% moisture.

Fertilizer Value of Poultry Manure*

			Pounds per Ton	
	Moisture %	Nitrogen	Phosphorous	Potash
Fresh Manure	75	29	10	8
Stored Manure	63.9	24	13	16
Broiler Litter	18.9	72	25	30
Layer litter	22.1	50	23	36
Slurry (Liquid)	92.0	22	12	7

When nitrogen is worth 25¢ per pound, phosphorous 15¢ per pound and potash 8¢ per pound the values per ton of poultry manure are:

Fresh Manure	$9.39
Stored Manure	9.23
Broiler Litter	24.15
Layer Litter	14.86

Other elements of plant food contained in poultry manure include calcium, magnesium, copper, manganese, zinc, chlorine, sulphur, and boron.

Based on information from Univ. of New Hampshire *Bulletin 444 — Farm Manure — 1966.*

HANDY WEIGHTS AND MEASURES TABLES

3 teaspoons = 1 tablespoon
2 tablesppons = 1 fluid ounce = 6 teaspoons
4 tablespoons = 12 teaspoons = ¼ cup = 2 fluid ounces
1 cup = 16 tablesppons = 8 fluid ounces
2 cups = 32 tablespoons = 1 pint = 16 fluid ounces
2 pints = 64 tablespoons = 1 quart = 4 level cups
4 quarts = 8 pints = 1 gallon = 16 level cups
16 ounces = 1 pound
6 tablespoons (level) = approx. 1 oz. of dry weight (for Wettable Powder only)

1 kilogram (kg) = 1000 grams (g) = 2.2 lbs.
1 gram (g) = 1000 milligrams (mg) = .35 ounce
1 liter = 1000 milliliters (ml) or cubic centimeters (cc) = 1.058 qts.
1 milliliter or cubic centimeter = .034 fluid ounces
1 milliliter or cubic centimeter of water weights 1 gram
1 liter of water weighs 1 kilogram
1 lb. = 453.6 grams
1 ounce = 28.35 grams
1 pint of water weighs approximately 1 lb.
1 gallon of water weighs approximately 8.34 lbs.
1 gallon = 4 qts. = 3.785 liters
1 quart = 2 pints = .946 liters
1 pint = .473 liters
1 fluid ounce = 29.6 milliliters or cubic centimeters

1 part per million (ppm) = 1 milligram/liter
                     = 1 milligram/kilogram
                     = .001 per cent
                     = .013 ounces in 100 gallons of water

1 percent = 10,000 ppm
            = 10 grams per liter
            = 10 grams per kilogram
            = 1.33 ounces by weight per gallon of water
            = 8.34 pounds/100 gallons of water

.1   percent = 1000 ppm = 1000 milligrams/liter
.01   percent = 100 ppm = 100 milligrams/liter
.001 percent =   10 ppm =   10 milligrams/liter
.0001 percent =   1 ppm =   milligram/liter

CUBIC MEASURE

cu. ft. = 1728 cu. in.
cu. ft. = 7.48 gals.
cu. ft. = .037 yd.
cu. ft. = 0.0804 bushels

cu. yd. = 27 cu. ft.
cu. yd. = 46,656 cu. in.

cu. yd. = 202 gals.
cu. yd. = 21.71 bushels
gallon = 269 cu. in (dry)
gallon = 231 cu. in (liquid)
gallon = 0.134 cu. ft.
lb. water = 27.68 cu. in.

lb. water = .016 cu. ft.

Glossary of Terms

Abdomen   Area between the keel and pubic bones.

Air Cell   Air space usually found in the large end of the egg.

Albumen   The white of the egg.

American Breeds   Those breeds developed in America and having common characteristics such as, yellow skin, non-feathered shanks, red earlobes. All lay brown eggs except the Lamonas which produce white eggs.

Axial Feather   The short wing feather between the primaries and secondaries.

Baby Chick   Newly hatched chick before it has been fed or watered.

Bantam   Diminutive fowl. Some are distinct breeds, others are miniatures of large breeds.

Beak   Upper and lower mandibles of chickens, turkeys, phesants, peafowl, etc.

Bean   Hard protuberance on the upper mandible of waterfowl.

Bill   The upper and lower mandibles of waterfowl.

Blood Spot   Blood in an egg caused by a rupture of small blood vessels usually at the time of ovulation.

Breast   The forward part of the body between the neck and the keel bone.

Breast Blister   Enlarged discolored area or sore in the area of the keel bone.

Breed   A group of fowl related by ancestry and breeding true to certain characteristics such as body shape and size.

Broiler-Fryer   Young chickens under 12 weeks of age of either sex, that are tender-meated with soft, pliable smooth-textured skin and flexible breastbone cartilage.

Brooder   Heat source for starting young birds.

Broody   Maternal instinct causing the female to set or want to hatch eggs.

Candle   To determine the interior quality of an egg through the use of a special light in a dark room.

Cannibalism   The habit of eating other birds in the flock.

Capon   A castrated male fowl having undeveloped comb and wattles, and longer hackles, saddle and tail feathers than the normal male.

Chalazae   White, twisted rope-like structures which tend to anchor the egg yolk in the center of the egg, by their attachment to the layers of thick albumen.

Chalaziferous layer   Thin layer of thick white surrounding the yolk and is continuous with the chalazae.

Cloaca   The common chamber or receptacle for the digestive, urinary and reproductive systems.

Cock   A male bird over 12 months of age.

Cockerel   A male bird under 12 months of age.

Coccidiostat   A drug used to control or treat coccidiosis disease.

Comb   The fleshy prominence on the top of the head of fowl.

Crop   An enlargement of the gullet where food is stored and prepared for digestion.

Crossbred   The first generation resulting from crossing two different breeds or varieties or the crossing of first generation stock with first generation stocks resulting from the crossing of other breeds and varieties.

Cull   A bird not suitable to be in a laying or breeding pen or not suitable as a market bird.

Culling   The act of removing unsuitable birds from the flock.

Debeak   To remove a part of the beak to prevent feather pulling or cannibalism.

Drake   Male duck.

Dub   To trim the comb and wattles close to the head.

Duck   Any member of the family Anatidae and specifically a female.

Duckling   The young of the family Anatidae.

Ear Lobe   Fleshy patch of skin below ear. It may be red, white, blue or purple, depending upon the breed of chicken.

Embryo   A young organism in the early stages of development, as before hatching from the egg.

Face   Skin around and below the eyes.

Flight Feathers   Primary feathers of the wing, sometimes used to denote the primaries and secondaries.

Follicle   Thin, highly vascular ovarian tissue containing the growing ovum.

Fowl   Term applied collectively to chickens, ducks, geese, etc. or the market class designation for old laying birds.

Gander   Male goose.

Germinal Disc or Blastodisc   Site of fertilization on the egg yolk.

Gizzard   Muscular stomach. Its main function is grinding food and partial digestion of proteins.

Goose   The female goose as distinguished from the gander.

Gosling   A young goose of either sex.

Gullet or Esophagus   The tubular structure leading from the mouth to the glandular stomach.

Hackle   Plumage on the side and rear of the neck of fowl.

Hen   A female fowl more than 12 months of age.

Hock   The joint of the leg between the lower thigh and the shank.

Horn   Term used to describe various color shadings in the beak of some breeds of fowl such as the Rhode Island Red.

Hover   Canopy used on brooder stoves to hold heat near the floor when brooding young stock.

Infundibulum or Funnel   Part of the oviduct which receives the ova (egg yolks) when they are ovulated. Fertilization of the egg takes place in the infundibulum.

Isthmus   Part of the oviduct where the shell membranes are added during egg formation.

Keel Bone   Breast bone or sternum.

Litter   Soft, absorbent material used to cover floors of poultry houses.

Magnum   Part of the oviduct which secrets the thick albumen or white during the process of egg formation.

Mandible   The upper or lower bony portion of the beak.

Meat Spots   Generally blood spots which have changed in color due to chemical action.

Molt   To shed old feathers and regrow new ones.

Oil Sac or Uropygial Gland   Large oil gland on the back at the base of the tail — used to preen or condition the feathers.

Ova   Round bodies (yolks) attached to the ovary. These drop into oviduct and become yolk of the egg.

Oviduct   Long glandular tube where egg formation takes place and leading from the ovary to the cloaca. It is made up of the funnel, magnum, isthmus, uterus and vagina.

Pendulous crop   Crop that is usually impacted and enlarged and hangs down in an abnormal manner.

Plumage   The feathers making up the outer covering of fowls.

Poult   A young turkey.

Poultry   A term designating those species of birds which are used by man for food and fiber and can be reproduced under his care. The term includes chickens, turkeys, ducks, waterfowl, pheasants, pigeons, peafowl, guineas and ostriches.

Primaries   The long stiff flight feathers at the outer tip of the wing.

Pubic Bones   The thin terminal portion of the hip bones that form part of the pelvis. Are used as an aid in judging productivity of laying birds.

Pullet   Female chicken less than 1 year of age.

Recycle, or force molt   To force into a molt with a cessation of egg production.

Relative Humidity   The percentage of moisture saturation of the air.

Replacements   Term used to describe young birds which will replace an old flock.

Roasters   Young chickens of either sex, usually 3 to 5 months of age, that are tender-meated with soft, pliable, smooth textured skin and with a breastbone cartilage, somewhat less flexible than the broiler-fryer.

Rodent   Any one of several gnawing mammals such as rats, mice and squirrels.

Roost   A perch on which fowl rest or sleep.

Secondaries   The large wing feathers adjacent to the body, visible when the wing is folded or extended.

Sex-linked   Any inherited factor linked to the sex chromosomes of either parent. Plumage color differences between the male and female progeny of some crosses is an example of sex-linkage. Useful in day old sexing of chicks.

Shell membranes   The two membranes attached to the inner egg shell. They normally separate at the large end of the egg to form air cells.

Slips   A male from which all of both testicles were not removed during the caponizing operation.

Snood   Fleshy appendage on the head of a turkey.

Sperm or spermatozoa   The male reproductive cells capable of fertilizing the ova.

Spur   The stiff, horny process on the legs of some birds. Found on the inner side of the shanks.

Standard-bred   Conforming to the description of a given breed and variety as described in the American Standard of Perfection.

Sternum   The breast bone or keel.

Stigma   The suture line or non-vascular area where the follicle ruptures when the mature ovum is dropped.

Strain   Fowl of any breed usually with a given breeder's name and which has been reproduced by closed-flock breeding for 5 generations or more.

Testicles or testes   The male sex glands.

Tom   A male turkey.

Trachea or Windpipe   That part of the respiratory system that conveys air from the larynx to the bronchi and to the lungs.

Undercolor   Color of the downy part of the plumage.

Uterus   The portion of the oviduct where the thin white, the shell and shell pigment are added during egg formation.

Vagina   Section of the oviduct which holds the formed egg until it is laid

Variety   A sub-division of breed — distinguished either by color, color and pattern or comb type.

Vent or Anus   Is the external opening from the cloaca.

Viability   The state of being able to live and grow.

Vitelline membrane   Thin membrane which encloses the ovum.

Wattles   The thin pendant appendages at either side of the base of the beak and upper throat, usually much larger in males than in females.

Windpuffs   Air trapped under the outer skin as a result of rupturing the air sacs during caponization.

Xanthophyll   One of the yellow pigments found in green plants, yellow corn, fatty tissues and egg yolks.

Yolk   Ovum, the yellow portion of the egg.

Index